MAN-MADE DISASTERS

BARRY A. TURNER
University of Exeter

WYKEHAM PUBLICATIONS (LONDON) LTD
(A member of the Taylor & Francis Group)
LONDON 1978

First published 1978 by Wykeham Publications (London) Ltd.

© 1978 B. A. Turner
All rights reserved. No part of this publication may be reproduced, stored in a retrieval system, or transmitted, in any form or by any means, electronic, mechanical, photocopying, recording or otherwise, without the prior permission of the copyright owner.

ISBN 0 85109 750 2 (Paper)

Text set in 11/12pt Photon Times, printed by photolithography and bound in Great Britain at The Pitman Press, Bath

Distribution

UNITED STATES OF AMERICA AND CANADA
Crane, Russak & Company, Inc.,
347 Madison Avenue, New York, N.Y. 10017, U.S.A.

AUSTRALIA, NEW ZEALAND AND FAR EAST
Australia and New Zealand Book Co. Pty. Ltd.,
P.O. Box 459, Brookvale, N.S.W. 2100.

JAPAN
Kinokuniya Book-Store Co. Ltd.,
17-7 Shinjuku 3 Chome, Shinjuku-ku, Tokyo 160-91, Japan.

INDIA, BANGLADESH, SRI LANKA AND BURMA
Arnold-Heinemann Publishers (India) Pvt. Ltd.,
AB-9, First Floor, Safdarjang Enclave, New Delhi-110016.

ALL OTHER TERRITORIES
Taylor & Francis Ltd., 10-14 Macklin Street, London WC2B 5NF

Contents

1. The need to understand disasters — 1
2. Attempts to understand disasters—I — 8
3. Attempts to understand disasters—II — 33
4. Three disasters analysed — 49
5. The incubation of disasters — 81
6. Errors and communication difficulties — 99
7. Order and disruption — 126
8. Information, surprise and disaster — 138
9. Disasters and rationality in organizations — 160
10. The origins of disaster — 189

Appendix — 202

Notes — 206

Index — 240

To my mother and father

1. The need to understand disasters

Mankind lives with the knowledge that its plans and intentions may be thwarted by disaster. Even when it thinks itself safest, it can never wholly discount the possibility that some unforeseen and destructive event may subject men to danger, or disrupt their orderly everyday affairs. Such events have commonly been regarded as resulting from some external and unfathomable force which could not be directly controlled, but only accepted. Traditionally, this force was seen as the outcome of a divine caprice, or as an expression of divine judgement; though, with the coming of the scientific age, an alternative view regarded it as the consequence of random shifts in the forces of a universe governed by chance.

But since the industrial revolution, these assumptions that man becomes involved in disasters only as an unwilling victim have become increasingly untenable. It is gradually becoming clear that many disasters and large-scale accidents display similar features and characteristics, so that the possibility of gaining a greater understanding of these disturbing events is presented to us. There is a purely intellectual interest, of course, in pursuing such understanding, but this is not merely an intellectual topic. There is an important and pressing need to gain a greater knowledge of the circumstances which precede and surround disasters, for several related reasons. First, the increasing size of the world's population, and the tendency of this growing population to concentrate itself in major centres increases the likelihood that any natural event such as a hurricane, a flood or an earthquake will adversely affect a large number of people. Secondly, the sources of energy which men control, and which possess the potential for the creation of man-made disasters, are coming under the authority of centralized bodies and organizations, and are thus increasingly vulnerable to misuse if major errors are made at the centre. Thirdly, the kinds of energy which man now makes use of are inherently much more destructive than those which he has traditionally controlled. Fire as a destructive force is still with us, but it has now been supplemented by the potentially damaging properties of modern high explosives, by the range of materials manufactured in chemical plants, from chlorine gas and dioxins to poisons like polybrominated biphenyl

(PBB), and by the stocks of radioactive materials being manufactured and stored by the nuclear engineering industry. Finally, man has begun during the twentieth century to intervene more frequently and on a larger scale in the processes of the environment which supports him, so that the possibility that he may upset some balance of natural forces to provoke a disaster becomes a very real one.

Taking all of these factors together, it does not seem surprising that there should be an upsurge of interest in gaining an understanding of disasters which will go beyond the knowledge provided by sensational accounts of such events. The similarities displayed by many past disasters can no longer be ignored, for future disasters threaten to be too extensive, too destructive and too cataclysmic to make it sensible to devote time and energy only to the problems of 'mopping up' after the event. The time has come to set aside the assumption which has been tacitly made for so long, that all disasters are unique, and that they are caused by singular chains of events which are beyond the reach of rational consideration because they are not amenable to generalization.

We have gained an expertise in handling and reducing hazards from many of the activities of our industrial society, but we are constantly proposing new projects and new kinds of intervention in the world. The advocates of nuclear power press the case for increasing our commitment to this energy source, and the scientists working with DNA are eager to extend their experiments with the 'stuff of life', but when there is a need to make decisions about such issues, many of us feel a sense of unease, wondering whether we are opening up a new Pandora's box of troubles and disasters for the future.

Even when such doubts are openly expressed, however, it is difficult to be sure that the debates that we undertake tackle the issues in the most appropriate manner. There are two obstacles to the intelligent assessment of such problems. The first is the lack of any widespread understanding of the way in which the potential benefits and dangers of a proposed course of action can best be estimated for decision-makers. The second obstacle is the lack of any kind of general framework which can help us to organize our understanding of the kinds of event which combine to create disasters, for without such a means of understanding past calamities, we are handicapped in trying to cope with the future.

Some contributions have been made in recent years to the removal of the first obstacle, and we now have at least two informed discussions of the problems of assessing risk and of reaching reasonable conclusions about what is acceptable or unacceptable.[1]* The second area of

* References are listed by chapter at the end of this book, pp. 206–239.

difficulty has received less attention, though it is intimately connected with the first, and the present study is intended to offer, not a way of preventing disasters—for this, of course, would presume a divine status—but a way of formulating general rules and principles about the emergence of such events, derived from an examination of evidence available to us from past disasters and large-scale accidents.

In looking for some general principles which may help us to understand the creation of disasters, however, we should not restrict our attention to purely technical causes, for organizational and other social issues are also likely to be involved. It is better to think of the problem of understanding disasters as a 'socio-technical' problem, with social, organizational and technical processes interacting to produce the phenomena to be studied. The present book begins with this assumption, and aspects of both social and technical factors leading to disaster will be examined, though the emphasis will fall upon the more neglected social elements.

In the course of this examination, we shall argue that an improved understanding of the nature of disasters can only be achieved by realizing that disasters arise from an absence of some kind of knowledge at some point. Questions about who had foreknowledge of an event, and who failed to predict it, about who was aware of a potentially dangerous situation and who was not are thus of central importance if disasters are considered from a socio-technical point of view. Consequently, one of the aims of this book is to demonstrate how an understanding of a number of aspects of disaster can be gained by seeking to discover the way in which knowledge and information relating to events provoking a disaster were distributed before the disaster.

In examining this 'social distribution of knowledge' about hazards, however, we are not merely concerned with looking for patterns of prophecy, with locating those who foretell impending doom. The knowledge with which we shall be concerned is less dramatic and more pervasive. Disasters may arise because men have insufficient knowledge of those natural forces which they try to harness, so that energy is released at the wrong time, at the wrong rate or in the wrong place. Therefore, it is crucial to consider the processes by which man gathers, discharges and uses energy, and the way in which he acquires, distributes and controls information relating to these processes. Evidently, this is not merely a matter of examining the technical information available to engineers and scientists, for, as succeeding chapters will demonstrate, many disasters arise solely from administrative causes, or from a combination of technical and administrative causes. Those in positions of power, those concerned with management and decision-making and those who control administrative machinery may well find that some of their actions contribute inadvertently to the causes of a disaster. Since, as I have already suggested, many of the major accumulations of energy which are likely to generate disastrous outcomes are the responsibility of large organizations, and of those who run them, the administrative component of disaster is one which it is important to consider. An aspect of this administrative involvement which is of particular interest concerns the interrelationship which exists between power and officially approved knowledge. The influence and approval of those in positions of authority may have a considerable effect upon the perceptions of the environment, and the theories about the working of the environment held by those in subordinate positions, and at times this constraint may have gross adverse effects for the organization concerned.

The scale of disasters which modern technology makes possible, and the increased likelihood of damage which is engendered by growing concentrations of organizational power and denser concentrations of population would make a compelling enough case for attempting to gain a better understanding of disasters. But a further consideration which prompted the investigation set out in this book means that the arguments presented should be of importance not only to those who are professionally concerned with safety and hazard control, but also to those who are more generally interested in the processes of administration. The conditions under which large organizations or groups of organizations come to make miscalculations which lead to large-scale physical disasters represent particular instances of a more general case. In gaining a better understanding of the way in which errors resulting in death and disruption arise, we may also be extending our knowledge of the way in which equally important but less physically destructive failures come about. Organization is centrally concerned with intention and the execution of intention. Disasters always represent failures of intention, and if we can come to know in more detail how such failures occur, we may also learn lessons which can be applied to organizational planning in general.

All disasters may be regarded as the outcomes of misdirected energy, so that the origins of disaster must be sought in the circumstances which enable or permit energy to be misdirected so as to transform the world in undesirable ways. From the point of view of many individuals and groups concerned to cope with the after-effects of violent upheavals, there are no crucial differences between intentional and unintentional energy discharges, for the medical problems of dealing with injuries, the civil problems of rescue, aid and recuperation, and the human problems of dealing with shock, trauma and disruption are essentially similar in both cases. But an examination of the origins of disaster cannot ignore this distinction, for if there are problems about the origins of an act of war or an act of terrorism, these are problems for the diplomat, the strategist and the political scientist. These problems are quite different from those of understanding the reasons for inadvertent energy releases, or the sources of ignorance or poor intelligence about future hazardous events.

In one sense, these latter problems pose questions for the engineer, or for the communications scientist, but the answers to these questions have been known in general terms to such specialists for a long time: build better, safer means of control, and improve communications about hazards and safety procedures. What is now needed is the rephrasing of these questions in a sociological and psychological form so that research questions may be asked not just about the technical expertise and practices which surround the prediction and regulation of energy discharges, but about the social context within which such practices develop and are carried out.

Western civilization places a high value upon rationality, and this civilization is sustained, if not dominated, by clusters of organizations, large and small, which profess an intention to display this value by pursuing rational courses of action. In such a society, the occurrence of a disaster indicates that there has been a failure of the rational mode of thought and action which is being relied upon to control the world, a threatening failure; and in seeking to understand such events, we must expect to make some examination of the nature of the rational model, and of the organizations and institutional settings in which it is prevalent.

The organizations which pervade modern society have become more dominant and far-reaching in their activities over the past two hundred years, so that more aspects of individual and societal life are subject to influence by organizational decision-makers than at any previous time. This extension of organizational interests has still greater significance for an investigation into the nature and origins of disasters because of the processes of centralization and concentration, and the increasing growth of organizational interdependence.

In most sectors of organizational activity, the tendencies towards centralization and domination by a small number of the largest organizations appear to proceed inexorably. But the large and dominant organizations do not enjoy complete freedom of action, because they are increasingly required to take account of the properties of the environment in which they operate. They are not functioning in the environment of 'perfect competition' which some early economists put forward as a theoretical ideal, and for this reason they must take account of the actions, policies and intentions of organizations in their own sector, in other sectors and in government. They cannot assume that the environment in which they are operating will remain more or less the same, for many of the changes which they now encounter are quite turbulent ones [2], and they have to accept that the modern organizational environment is often characterized by unpredictable, large-scale disturbances [3]. In such an environment, it becomes much less possible to assume that the costs of unintended outcomes will be borne by the environment.

The increasing dominance of small numbers of large organizations in many areas of society produces conditions under which society as a whole becomes more dependent upon 'correct' decisions about energy and hazards being made by those few executives with national and international responsibilities. Society gains the benefits of standardization and uniformity of action by sacrificing the opportunity to generate a variety of options, in situations in which we do not necessarily know which option will be the best.

When a small organization operating under conditions of perfect competition makes a seriously false assumption, it goes out of business, with only minor consequences for society, but when a large organization in a monopolistic or oligopolistic position makes a similar error, the consequences will be much more far-reaching. Large organizations in industrial society have devoted considerable time and energy to improving the effectiveness with which they implement decisions, but it is not clear that the effectiveness of self-correcting mechanisms within such societies has improved to the same degree. It is important, both for organizations and for governments in modern society to make sure that the 'intelligence' by which they guide and correct their actions is of as high a quality as possible. High-quality intelligence should be clear, timely, reliable, valid, adequate and wide-ranging, and persuasive arguments have been presented for the development of research into ways of improving the quality of intelligence available to our organizations [4].

Modern organizations are concerned to operate and to implement decisions using and promoting modern technology, the technology which has wrought such an unprecedented change over virtually every aspect of human activity over the past two hundred years. The myth of Prometheus may be read as conveying the message that the increased use of technology cannot be pursued without increasing the stakes in the game which man plays with nature, and modern technology, beneficial though many of its aspects may be, is concerned with the handling of more concentrated packets of energy than ever before. With the use of this energy to produce forms of matter and patterns of activity which are novel and potentially destructive, the capacity for error is accordingly increased.

As organizations grow and spread, and as many associated cultural shifts take place, society comes to be increasingly predicated upon assumptions of control. Our dependence upon large-scale organizations for so many of the essentials of life makes it crucial that patterns of organizational control continue to function. Not only do we need to be able to execute those programmes which we plan for ourselves; we also need to be able to understand our environment sufficiently well to gain an accurate grasp of the likely effects of our actions upon it. Because the model of rational intelligence-gathering, anticipation and control is so central to Western man's view of his relationship to the world, it was suggested above that major failures of control constitute threats to this relationship, even where a cataclysmic outcome is avoided. For this reason, an investigation of the nature and origins of disaster must expect to gather much of its evidence from those sensitive and crucial marginal areas where intendedly rational action fails.

2. Attempts to understand disasters—I

If we are to try to gain an understanding of disasters, it would be foolhardy to plunge ahead without first of all considering the advice, knowledge and experience of others who have studied this topic; but, unfortunately, the guidance offered by previous workers in the field is less clear than might have been hoped. The problem is not one of a paucity of material, for many accounts and discussions of disasters are available. Rather, it is a matter of finding investigators who are concerned to gain an understanding of disasters from the point of view set out in the preceding chapter, and investigators who approach the topic with a concern for the origins of disaster seem to be very rare indeed. Most of those who have written about disasters have had their attention focused upon rather different issues, and their analyses do not, therefore, address themselves directly to our problems. As a consequence, we shall need to identify elements in the range of rather disparate approaches previously followed which may be taken as relevant to the development of an understanding of the socio-technical contexts within which disasters develop.

2.1. Accounts, Classifications and Lists of Disasters

First-hand accounts of disasters are of little direct use for our purposes, for, from Pliny onwards, they are accounts of devastating or disruptive events from the points of view of victims or near victims; whilst they may provide much useful information about human emotions and reactions in moments of severe stress, they generally provide little detailed or accurate information about the origins and nature of the event which is producing the distress. In the case of disasters, the normal problems of obtaining accurate and reliable accounts of events from untrained observers are complicated by the distortions which such stressful events place upon memory and upon perception, and by the deeply felt emotions aroused by such events [1]. Journalistic accounts of disasters often suffer from similar deficiencies and inaccuracies, and have been criticized as inadequate sources of data for those concerned to investigate and understand disasters [2].

ATTEMPTS TO UNDERSTAND DISASTERS—1

There is, of course, a fascination about large-scale destruction which leads sightseers to crowd towards disaster areas [3], and which also ensures that there will be a ready readership for accounts of catastrophes. Whether this interest arises from a genuine concern with the practical problems of coping with the disruption, from a curiosity about how one would fare oneself under such misfortune, from a desire to examine the limits of the normal categories of experience [4], or from deeper urges of the psyche [5], it is likely to colour those accounts which are written with a wider audience in mind, adjusting the accounts to meet the requirements of the public which also buys disaster novels and pays to see disaster films [6].

The major limitation of first-hand accounts of disasters for our present purpose, however, is the inadequate discussion which they provide of the sources and origins of the events described. For this reason, when empirical material is used later in this book, it will be drawn from official reports and discussions, a little more removed from the events than the immediate accounts of survivors, although such sources, set out in summary form in Table 2.1, have their own deficiencies and limitations.

Many writers have attempted to understand the phenomenon of disaster by compiling lists of disasters which specify the area of concern, and thus set bounds to the problem. A list of 294 'notable' famines outside India, and 80 famines inside India has been compiled by Keys [7], for example, and a number of other writers have produced similar listings of particular types of disaster, or summaries of incidents for particular regions [8]. Lists are also available for more specialized categories of incidents such as British railway accidents [9], and major British mining disasters [10].

The range of attempts made to delimit the area fails to provide any fully satisfactory account of the nature and extent of those incidents which might be regarded as disasters, presumably because of the daunting magnitude of the task and because of the difficulties of definition. It does not seem to have been possible to identify wholly satisfactory criteria for inclusion, except in the most specialized lists, and the amorphous nature and sprawling distribution of many of the events referred to make it doubtful that the more general lists are at all complete. Most of the lists exclude at least one or two of the more important types of disaster; even where the lists are concerned with similar kinds of event, incompatibilities exist between them, as with the 300 000 casualty famine listed by Montandon for China in 1850 which is completely omitted by Keys [11].

An alternative approach, which tries to limit the numbers of disasters to be accounted for by dealing only with 'serious' disasters with more

TABLE 2.1. *Summary of Western's classification of sources of accounts of disasters and some of his main comments.*

Classification	Significant sources	Comments
I. Press reports		Often inaccurate in many respects
II. Non-fictional literary accounts	Merewether (1898) Fisher (1927) Hersey (1946) Masters (1950) Furneaux (1964) Roueché (1967) Salisbury (1969)	
III. National Reports		
A. Government sources	Western discusses USA, Canada and Japan, but omits the United Kingdom	Government reports are not common in all countries because central government is not informed where information-handling is decentralized; because no report is thought appropriate if a disaster is thought of as an internal problem; and because resources for preparation of reports are often lacking.
B. Non-government sources	1. National Red Cross 2. Disaster Research Center, Ohio, USA 3. Department of Geography, Toronto Canada. 4. Office of Information, NATO 5. Disaster Research Centre, Bradford University, UK.	Now holds the National Academy of Sciences/National Research Center disaster files. Trying to bridge the gap between the natural and social science approach in its working papers: Natural Hazard Research (1971); Kates (1970).
IV. International A. International agency reports (UNO, WHO, etc.)	Ambraseys (1968) Robson and Canales (1968) Arsovski (1970) UNESCO (1971) Pias and Stuckmann (1970) Mackey *et al.* (1971)	UNESCO teams reporting on earthquakes, floods, typhoons, etc., mainly with geophysical concerns.
B. International relief agencies	International Relief Union journal, published quarterly as: *Matériaux pour l'étude des calamités*, 1924–37; as *Revue pour l'étude des calamités*	Chronicle of natural disasters. Patterns and effects discussed from a geophysical point of view. Medical and social aspects usually restricted to number of dead and injured, and number of structures destroyed.

TABLE 2.1.—(Cont'd)

	1938–62; and as *Revue de l'Union Internationale de Secours*, 1963—present.	
V. Scientific and technical reports A. Natural science journals	1. *Geographic Reviews* (New York City) 2. International Relief Union journal. 3. *The Lancet* 4. *Mass Emergencies*	See above. A new interdisciplinary journal, first published October 1975.
B. Sociological studies	Prince (1920) US Strategic Bombing Survey (1947) Oughterson and Warren (1956) Killian (1953) *Unscheduled Events*	Quarterly restricted circulation journal of the Ohio Disaster Research Center.
C. Medical studies	Blocker and Blocker (1949) Saidi (1963) *Maroc. Med.* (1961) *Vojno-Sanit. Pregl.* (1964) *Public Health Report* (1964) *Munch. Med. Wschr.* (1962)	Peak of interest during and after Second World War, but also includes reports of medical aspects of disaster such as earthquakes in Iran and Skopje.

Source: Compiled, with additions, from Western's study. See note 2.

References

Ambraseys, N. N., Zatopek, A., Tasdemiroglu, M. and Aytun, A. (June 1968) *Turkey: the Mudurnu Valley (West Anatolia) Earthquake of 22 July 1967* UNESCO Report 622 (Paris: UNESCO).
Arsovski, M., Bouwkamp, J. Cismigiu, A. and Izumi, M. (June 1970) *Yugoslavia: The Banja Duka Earthquakes of 26 and 27 October 1969* UNESCO Report 1919 (Paris: UNESCO).
Blocker, V. and Blocker, T. G. (1949) The Texas City disaster: a survey of 3000 casualties, *American Journal of Surgery* **78,** 756.
Fisher, H. H. (1927) *The Famine in Soviet Russia, 1919–1923* (London: Macmillan).
Furneaux R. (1964) *Krakatoa* (Englewood Cliffs, NJ: Prentice Hall).
Hersey, J. (1946) *Hiroshima* (New York: Knopf).
Kates, W. (1970) *Natural hazard in human ecological perspective: hypotheses and models* Working Paper 14. (University of Toronto, Canada: Natural Hazard Research).
Killian, L. M. (1953) *A study of response to the Houston, Texas, Fireworks Explosion* Publication 391 (Washington, DC: National Academy of Sciences/National Research Council).
Mackey, S. Finney, C. and Okubo, T. (May 1971) *The Typhoons of October and November, 1970* UNESCO Report 2387 (Paris: UNESCO).

TABLE 2.1.—(Cont'd)

Maroc. Medical Abbes, Y. B. (1961) Agadir: lessons of an earthquake, *Maroc. Medical* (Casablanca) **40** (No. 2) pp. 111–112. Introduction to special issue on the Agadir Earthquake.
Masters, R. V. (1950) *Going to Blazes* (New York: Sterling).
Merewether, F. H. S. (1898) *A tour through the famine districts of India: first person account by Reuter's special correspondent* (London: Innes and Co.).
Munch. Med. Wschr. (19 Oct. 1962), Bewältigung und Lehre der Hamburger Flutkatastrophe vom Februar 1962 (A collection of articles on the 1962 Hamburg floods) *Münchener Medizinische Wochenschrifte* **104** (No. 42) pp. 1973–2001.
Natural Hazard Research (1971): *summary of a series of working papers 1968–1971* (Canada: University of Toronto).
Oughterson A. W. and Warren, S. (1956) *Medical Effects of the Atomic Bomb in Nagasaki* (New York: McGraw-Hill).
Pias, J. and Stuckmann, G. (June 1970) *Tunisie: les inondations de septembre–octobre, 1969, en Tunisie* UNESCO Report 1957 (Paris: UNESCO).
Prince, S. H. (1920) *Catastrophe and Social Change: based upon a sociological study of the Halifax disaster* (New York: Columbia University Press).
Public Health Reports:
Wilson, M. R. (Oct. 1964) Effects of the Alaska earthquake on functions of PHS Hospital, pp. 853–861.
Parrish, H. M., Baker, A. S. and Bishop, F. M. (Oct. 1964), Epidemiology in public health planning for natural disasters, pp. 863–867.
Both in *Public Health Reports* (Washington, DC) **79** (No. 10).
Robson, G. R. and Canales, L. (May 1968) *Venezuela: the Caracas earthquake of 29 July 1967* UNESCO Report 571 (Paris: UNESCO).
Roueché, B. (1967) *Annals of Epidemiology* (Boston: Little, Brown).
Saidi, F. (25 Apr. 1963) The 1962 earthquake in Iran: some medical and social aspects, *New England Journal of Medicine* **286** (No. 17), 929–932.
Salisbury, H. E. (1969) *The 900 Days: the Siege of Leningrad* (London: Secker and Warburg).
UNESCO (1971) *Reunion consultative d'experts sur la seismicite du Maghreb, Tunis, April.1971* (Paris: UNESCO).
U.S. Strategic Bombing Survey (1947) *The Effects of Bombing on Health and Medical Services in Japan* Document 741801 (Washington, DC: U.S. Government Printing Office).
Vojno-Sanit. Pregl. (1964). 'Report' *Vojno Sanitetski Pregled* (Military–medical and Pharmaceutical Review—English edition) Belgrade **21**, 7–8.

than 100 000 dead also turns out eventually to be unsatisfactory, since there have been few disasters on such a scale until relatively recently in world history. Moreover, deaths are often as much a function of population density as of the nature of the disaster itself, and in any case the number of deaths may bear little relation to the long-term social, medical and economic effects of a disaster [12].

A related way of trying to bring some order to the study of disaster in a preliminary manner is by the construction of classifications which separate out different kinds of disasters. These attempts do provide a way of gaining some understanding of the kinds of phenomena which are being dealt with, in the early, exploratory stages of an inquiry, but ultimately they do not enlighten us very much because they are unable to produce a satisfactorily rigorous result. Western, an American epidemiologist, has usefully reviewed various classifications of disasters, mostly based upon origin or cause as a classification criterion; combining much of the work of earlier writers, he has set out what he calls an 'Etiologic Classification' of disasters which is reproduced in Table 2.2

[13]. Although this classification is ambitious, Western warns us that it is not intended to be complete, and he adds as a further reservation that:

> 'I do not find the separation of disasters into natural or man-made events very productive. At one level, a disaster becomes a disaster only when man and the environment he has created are affected. An avalanche in an uninhabited valley, or an earthquake in the Arctic are geophysical events, not disasters' [14].

TABLE 2.2. *Western's Etiologic Classification of Disasters.*

A. *Natural disasters*
  1. Natural phenomena beneath the earth's surface:
     (*a*) earthquakes
     (*b*) tsunamis
     (*c*) volcanic eruptions

  2. Natural phenomena of complex physical origin at the earth's surface:
     (*a*) landslides
     (*b*) avalanches

  3. Meteorological/hydrological phenomena:
     (*a*) windstorms (cyclones, typhoons, hurricanes)
     (*b*) tornadoes
     (*c*) hailstorms and snowstorms
     (*d*) sea surges
     (*e*) floods
     (*f*) droughts

  4. Biological phenomena:
     (*a*) locust swarms
     (*b*) epidemics of communicable diseases

B. *Man-made disasters*
  1. Caused by warfare:
     (*a*) conventional warfare, including siege and blockade
     (*b*) non-conventional warfare (nuclear, biological, chemical)

  2. Caused by accidents:
     (*a*) vehicular (planes, trains, ships, cars)
     (*b*) drowning
     (*c*) collapse of buildings and other structures
     (*d*) explosions
     (*e*) fires
     (*f*) biological
     (*g*) chemical, including poisonings by pesticides and pollution

Source: Western (1972) Table I, p. 6. For full reference see Notes.
Western's table is adapted from G. P. Gill (1963), Organizacion de la asistencia

sanitaria en las calamidas publicas, *Medna trop.* **39,** 459ff.; and from M. Fournier d'Albe (1970), Natural disasters, their study and prevention, *UNESCO Chron.* **16,** 195–207.

The multiple causes often associated with disasters make the 'natural'/'man-made' distinction even more difficult to uphold in any rigorous way; the distinction is further blurred, on the one hand, by the increases in world population, which mean that any given geophysical event is more likely to affect man, and on the other hand by man's increasing ability to modify his environment in a way which may provoke what have been traditionally regarded as 'natural' disasters. The production of earth tremors by the forces accumulated when large artificial lakes are created, and the possibility that cloud-seeding might generate or exacerbate floods offer instances of possible 'natural' yet 'man-made' disasters, and this category seems likely to increase in the future [15].

We may conclude from this brief discussion that the topic of disaster is not a simple one, and that no completely satisfactory listings or classifications of disasters exist. Indeed, when we consider the manner in which we use the term 'disaster', alongside the almost ubiquitous nature of slips, errors and accidents of various kinds, we may feel, perhaps, that no adequate listing or classification of disasters will ever be constructed [16].

It was suggested above that many first-hand accounts and many journalistic accounts of disasters are partial, distorted or sensational, but there is an important category of writing which should be exempted from these criticisms. When a competent journalist, or a team of journalists, is given the time and opportunity to combine persistence in chasing facts and extensive access to sources with concentration upon a single 'story', the outcome may provide an extremely useful dossier. This not only details the nature of the disastrous event itself, and the manner in which it affected the lives of those involved, but also reviews the evidence available about the sources and origins of the incident.

The account of the 1975 Paris DC 10 crash by the *Sunday Times* 'Insight' team provides an excellent recent example of the genre, one which is particularly useful in this case because of the absence of any single, authoritative public report on the incident [17]. This investigation, however, while providing an extremely valuable account of a major disaster, also illustrates very clearly the limitations of this kind of reporting for those concerned to understand, not a single disaster, but the nature of disasters in general. The journalist is professionally preoccupied with the particular: he is trained to ferret out to the best of his abilities just who said or did what, when, and to record this information in relation to other events. Even if he has the inclination, he rarely has the opportunity to attempt to develop generalizations which express a particular set of findings in terms which are sufficiently abstract to be applied to other cases. The DC 10 report is unusual in presenting some general comments, but these are restricted, as is appropriate for the task being undertaken, to the field of air safety, and are not developed more widely.

2.2. Medicine and Disaster

Turning now to another professional group with an occupational interest in disaster, we can ask whether we are likely to find insights which will aid our attempt to gain an understanding of the nature of disaster in the work of those concerned with medicine. Medical personnel are, after all, most intimately concerned with disasters, for they are called out to cope with the distasteful aftermaths of all kinds of violent eruptions, whether these arise from natural or man-made sources, from war, from rioting or from terrorism. Clearly, many of the most pressing problems which face medical personnel in such situations are not significantly affected by variations in the nature of the causes of injuries they have to deal with, except insofar as the presence, say, of an enemy may make the necessary treatment more difficult to carry out.

Following a disaster or major accident, the problems which immediately confront medical personnel are practical or 'logistical' problems of getting medical assistance or supplies to the scene, or of transferring the injured to hospitals or temporary medical-care sites. As they are assessing these problems, they also need to take into account the requirements imposed by the likely injuries from different kinds of accident: from fire or smoke, from impact in a traffic collision, from explosion or civil disturbance, or from nuclear incidents. Treatment may be further complicated by such issues as whether 'the examination of human wreckage is a valuable form of investigation' following an air accident [18].

Triage, or the sorting of casualties according to the type and severity of their injury, needs to take place at the scene of the incident and in the hospital; the design and implementation of systems for documenting the injuries diagnosed and treatment already given are not negligible problems, particularly when a hospital is called upon suddenly to handle a large number of casualties [19].

Thus, from a medical point of view, the onset of a disaster presents major problems of response, management and planning, in addition to the purely medical problems of treatment [20]. Amidst all of these immediate and pressing medical problems, it is perhaps not surprising, as Western has commented, that 'Sociologists have paid more attention to the psychological problems of survivors of disaster than have physicians' [21].

The kinds of problem listed above are all intensified enormously, of course, when the numbers of dead and injured are large, and the medical specialist may be faced with major demands for treatment and for the safeguarding of public health in disorganized conditions such as those associated with extensive flooding [22], or with the aftermath of a large-scale earthquake [23]. And even when the injuries of the survivors have been attended to, the medical specialist may find himself involved in a consideration, say, of the logistical problems of rapidly and effectively disposing of numerous corpses [24].

There are thus very compelling reasons for the focusing of medical attention upon the period after disaster. But recently Western, the epidemiologist already mentioned, has reviewed the field of medical approaches to disaster [25]. It is his concern to draw attention to the post-disaster bias in the medical literature, and to suggest that it is time for an epidemiological approach to be applied to disasters. The epidemiologist reviews the incidence and distribution of outbreaks of various types of disease and injury with a view to identifying the particular combinations of prior circumstances associated with their emergence, tracing the causes where possible, in order to assist him in devising some form of prevention. Western was able to identify only two studies in the medical field which had taken this approach to disasters; one reviewing much of the disaster literature considered below, but with a particular emphasis on medical aspects [26]; the other discussing the 1970 East Bengal cyclone from an epidemiological point of view [27].

Epidemiological studies of outbreaks of disease in the past have led to the medical specialists concerned identifying both medical and non-medical causes. The preventative steps suggested by investigations into, for example, cholera or malaria have necessitated both the study of the medical origins of the conditions in question and the advocacy of public health measures such as improved public hygiene, or the eradication of mosquito breeding grounds as part of the strategy to deal with the original problem. In the disaster field, however, it seems unlikely that the epidemiologist will find many purely *medical* preconditions; anyone heeding Western's criticisms will find himself involved in pursuing precisely the questions that are being raised in this book. From Western's emphasis of the need for an epidemiological approach to disasters, we may realize that the problematic area for those concerned to understand disasters lies in the period before the disaster breaks, in the aetiology or epidemiology of disasters, or (to use non-medical terms) in the preconditions which give rise to the disaster. Questions about preconditions seem likely to lead to the identification of technical, social, administrative and psychological features of the pre-disaster situation; at this point, the concerns of epidemiologist and social scientist wanting to understand disaster seem to merge.

2.3. *Failures in Physical Systems*

Moving on to consider other possible approaches to the understanding of disaster, and particularly of the preconditions of disaster, we find another professional group with its own preoccupation with aspects of disruption and unwanted failures. For the engineer, the important disasters are those which arise from the malfunctioning of a purposive man-made system. A concern with such malfunctioning and its avoidance is central to the work of engineers; as a result, their discussions of failures and disruption are carried out within a world of discourse which is very different from that of the medical specialist or the social scientist. Sometimes, the professional engineer is involved in investigations of and inquiries into the nature of the technical events leading to a disaster, particularly when the disaster involves, say, the failure of a bridge [28], or the breakdown of a chemical plant [29]. But he would, in any case, be concerned with the examination of these events, whether or not a disaster had ensued; for within the technical world there is a central and longstanding preoccupation with physical and technical causes, not only of disasters but of failures of any form of equipment and, indeed, with malfunctioning of any kind.

Writers in this field, of course, are subject to the temptation referred to above—to exploit the dramatic fascination which accounts of large-scale failures exert on their potential readership, including, in this case, not merely the general public but also the engineer or designer. He has to live with the knowledge that, in his everyday work, some error, miscalculation or misjudgement may lead to a major accident. Jacob Feld, for example, during a lifetime's career in the American construction industry, has assembled a fascinating collection of instances of construction failures, both serious and bizarre [30]. Feld offers us a rich assemblage of reports of bridges collapsing, of roofs falling in, of buildings falling down, slipping sideways or dropping into holes in the ground; he includes a description of the American military 'Big Dish' project which the US Navy embarked upon in 1948 and abandoned in

1962, having spent $63 million on the construction of the 600 ft diameter radio telescope before the impossibility of successfully completing a project of such a size became apparent [31]. But collections such as this cannot be much more than bedtime reading for construction engineers; quite apart from the absence of source references, Feld makes no attempt to understand and analyse the cases presented, merely ordering the anecdotes and press reports according to whether the failure occurred at, below or above ground, and then according to the material (steel, wood, concrete, etc.) which failed.

More serious attempts to tackle the problem of failure and to create an adequate conceptual framework for its discussion do exist, though. The emphasis on fault-finding and reliability which characterized the NASA space programme gave a considerable boost to such an endeavour [32], and the enormous potential hazards inherent in the nuclear engineering industry have put a high priority on discussing, locating and minimizing the possible occurrence of faults in this area, and on assuring the public that such activities are being diligently pursued [33]. Developments in these two advanced sectors of engineering have taken place alongside engineers' continuing concern with the need to tackle the wider problem of improving engineering reliability in general [34].

All engineering systems are consciously designed to achieve a goal or a set of goals, and within such systems, or within other related systems such as information or data-processing systems, disruptive or unplanned-for events are seen less as 'disasters' than as instances of malfunctioning or failure. Failures may then be defined and discussed in the following kind of language: 'Failure (is) . . . a characteristic of sub-systems, namely "not contributing to the goal fulfilment of the supersystem" ' [35]. Or, alternatively, failure is 'the termination of the ability of an item to perform its required function' [36].

Definitions of this kind must stress 'goals' and 'functions', for engineering and design failures can only be measured against the standards set by such notions. Moreover the emphasis upon goals, or upon some similar statement of intent, may also be applied usefully to the study of disasters; for when we talk of disaster, by implication we are relating the event to some desirable state of affairs which would obtain if the disaster had not intervened. Our intention was that things should continue as normal, and only after the event do we recognize those factors and events which made it impossible for this 'goal' to be achieved. This discussion of purposive intent in relation to disasters will be developed more fully in later chapters; for the moment, it can be seen that there is not necessarily a great divergence between engineering approaches to the problems of failure and non-engineering studies of disaster. This is not to say, though, that the matter is a simple one; for failures may arise, not merely because some component or sub-system is unable to function or to achieve its goal, but also, for example, because its goal may not have been devised as it ought to have been if some wider system, in turn, is to achieve *its* goal. The complexities created by such interconnections can be developed almost indefinitely.

In any particular case, or within any particular system, if the investigator wishes to identify and examine actual instances of failure, he has to equip himself in some way with a set of agreed criteria which he can use to ascertain whether a given incident represents a failure or not. This fact immediately highlights a problem which has not previously been evident in the study of disaster, for the recognition of a disaster appears to present no difficulties: its existence is readily indicated by news bulletins, by flurries of activity on the part of medical and emergency services, and by the occurrence of damage to property and death or injury to people. But, just as it will be demonstrated below that the circumstances under which warnings of impending disaster should be issued are not always readily discernible, so the engineering experience indicates the existence of a major perceptual element which must be present in the definition of any incident or event as a failure or as a disaster.

To identify a failure, the observer must compare the actual present state of affairs with what, according to some agreed criteria, *ought* to have been the case: the outcome must be set against the goal. What complicates this process of comparison is the fact that goals, even in engineering systems, and often in social systems, may be fuzzy, ill-defined or ambiguous. And even when this is not the case, the goal can only be considered in terms of a description (of what ought to have been) at least partly constructed by the observer [37], since the goal has not actually been attained. We can never produce a complete description of any physical or social system that will remove all doubt about the eventual desired state, though we can circumscribe and limit the area in which interpretation is required [38].

Because the unmistakable identification of a failure cannot be obtained without these elements of perception and judgement, the issue is sometimes avoided by a kind of sleight-of-hand in which the intended goal or base for comparison is shifted. In this process, which is akin to the psychological adjustment called 'rationalization', it is explained that the present outcome was what was required all along; thus instead of a failure, we have a non-failure. A similar, but not quite identical kind of adjustment may also occur because of what has been called 'the degree of forgiveness of the system' [39]. Lindquis (who coined this delightful phrase) points out that, because of the difficulty of precisely and unambiguously specifying future goals, most systems have some degree of 'forgiveness' built into them. At times, then, outcomes from components, sub-systems or lower levels in the system which appear to be unacceptable according to the formal criteria laid down may, in fact, be perfectly adequate for the purposes for which they were intended. Thus, the system may accept as non-failures, results which are failures in formal terms [40].

Non-engineers often seem to stand in awe of the practitioners of applied science and engineering, regarding them as inhabitants of a world where rationality reigns supreme, where alternative courses of action can be measured and rigorously compared and where science informs every decision. The difficulties in precisely and unambiguously identifying goals bring this view into question to some extent; but a closer examination of the processes of engineering design, and of the conditions which have given rise to engineering failures in the past, reveals some additional limitations under which engineering designers must operate. In fact, such designers are often required to work with strictly limited sets of data on the materials they are to use, their properties and their reliability. Information about the operating conditions under which many pieces of equipment will be expected to function may be even scarcer, so that estimates of the likely range of stresses and strains to be coped with may sometimes be little more than informed guesses. Even when information about all of the above matters is available, existing physical, chemical and engineering theory may not allow of an unequivocal solution to the designer's problems, so that the ability to produce successful engineering designs has been spoken of as having elements of 'witchcraft' inherent in it, with the consequence that, in this inexact field, the successful design engineer is 'a man beyond price' [41].

Examining failures of equipment and plant reported in the Proceedings of the Institution of Mechanical Engineers in the period from 1950 to 1975, it is found that many of the failures arose from designs which extended beyond the knowledge or experience of the designer. Others relied for their success upon the satisfactory operation of materials which, even when of high quality, may well show a remarkably wide scatter of performance life. In addition, such designs may have been called upon to operate under conditions which would have produced a further wide range of unknown variations and fluctuations of stress [42].

This is not to suggest that all engineering structures are bound to display severe malfunctions, or that it is foolish to place any degree of trust

ATTEMPTS TO UNDERSTAND DISASTERS—I 21

in their operations. Rather, it is to point out that the processes of engineering design are limited by the kinds of constraints which restrict all human activities, and that the theoretical guidelines by which they are steered are incomplete and may display an improvised character. This does not mean that the persistent application of thought, concern and control to an area in which equipment may fail in a dangerous manner will not produce results. The sharp and continuing decline in the number of boiler explosions and in their severity during the period from 1885 to the present day shown in Figure 2.1 presents a clear record of the successful improvement of known hazardous conditions [43], and the improvement of the railway signalling systems during the nineteenth and twentieth century may be taken as another index of the manner in which rationality may be successfully and progressively applied in the engineering field [44].

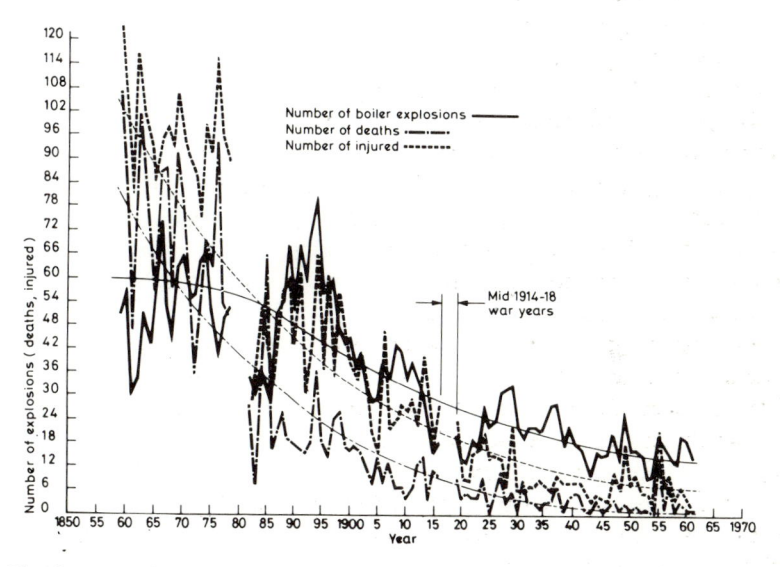

Fig. 2.1. Number of steam-boiler explosions between 1860 and 1960, together with the number of persons killed and injured. *Source:* J. E. Eyers and E. G. Nisbett 'Boilers' in R. R. Whyte (ed.) *Engineering Progress Through Trouble* (see note 34), p. 109 Figure 1. Reprinted by permission of the Council of the Institution of Mechanical Engineers.

But, notwithstanding such success stories, it is important, particularly in view of current developments in technology, and of the hazards associated with them, to realize the limits under which designers operate in their attempts to achieve their goals and to avoid failure. In the face of many problems, such as, for example, the design of gears or the design of bearings for large equipment [45], designers are often lacking in the necessary information upon which successful designs could be based, and, in a discussion of a series of instances of one particular type of bearing failure, three engineers specializing in this field have commented that such failures constitute: '... a lasting lesson to all engineers of the sort of trouble that can occur when any new departure from existing practice is made for the best possible reasons. The areas of our ignorance are still very large!' [46].

Many of these bearing failures were associated with the scaling up of existing satisfactory designs to operate in larger sizes and with larger forces, and similar processes of 'incremental' design have also led, for example, to failures in large-scale electricity generators. In the analysis of such a series of failures following on the use of a larger version of a tried and trusted design, which brought new factors into operation, it has been commented that although the basic reason for each of this series of failures appears to have been the same, the engineers concerned were 'rightly loath to change a design which has given many years of satisfaction' [47].

Other papers from the Proceedings of the Institution of Mechanical Engineers draw attention to an unsuspected 'size effect' which led, in combination with other marine engineering factors, to eleven ships breaking in half at sea in the two-year period 1967–69 [48]; and to the fatigue failure of a large power-station turbine shaft which was caused by design features apparently proved by nearly forty years of successful experience, but vitiated by increases in speed and stresses [49].

In the face of the often overwhelming problems of deciding which of thirty or forty features of a material specification are relevant to a particular design, and in the face of the impossibility of making realistic estimates of many of the stresses likely to be encountered by materials in operation, it may be asked how engineers cope with the problems of producing successful design. Many proven and clear-cut design guides exist and are made use of, of course, but in areas which they do not cover, the solutions to particular problems must be sought largely on a trial-and-error basis. In these cases, successful solutions may often precede theoretical explanations of them by many years [50], as in the case of the spiral fins which were found to solve the problem of preventing steel chimneys failing as a result of vibration [51]. It should be noted that where atheoretical solutions to problems are adopted, the possibility of predicting any resulting unintended consequences or long-run complications is lessened, since if it is not clear why a particular solution works, it is even more difficult to try to predict unwanted side effects [52].

The design process often makes use of numerical calculations in order to decide the sizes and types of materials to be used, but the indicators available for estimating the properties of materials in such calculations are often arbitrary or even irrelevant [53]. Alternatively, the prediction of likely defects in a particular design may have to be made by the use of nothing more sophisticated than 'rule-of-thumb' calculations [54]. Reinforcing the points made earlier about the need to specify an intended level of operation before an assessment of failures can be made, Birchon, an engineer concerned with maintaining quality control in ocean engineering, has pointed to some of the difficulties in specifying and maintaining a high level of quality [55]. Noting that no material is free from defects, with every cubic centimetre of good steel containing over 100 million defects, if displaced atoms are counted as such, Birchon points out that quality inspection cannot start until a level of unacceptable defects has been defined, and this in turn relies upon the anticipation of probable failure patterns, which is often a 'frighteningly complex' problem.

Issues like this are raised more clearly if an attempt is made to set out a clear rationale or procedure for choosing a factor of safety. The very large factors of safety commonly adopted in engineering may be separated, Nixon and his colleagues have pointed out, into two elements: a true safety margin, plus a margin of ignorance, which covers a wide variety of unknowns [56].

Ensuring the quality and durability of materials and individual components is not the only problem of engineering design reliability, however, for even if the information to design individual components accurately and safely were always available, problems of failures due to interaction effects between components would still remain. Whyte, in a discussion of this problem comments:

'There failure of a new and complex system is hardly ever due to the wrong design of one component: what is needed is the compatibility of many components, and this can usually be found only by long and strenuous development work' [57].

The significance which Whyte attaches to interaction effects is reinforced by an examination of, for instance, the causes of the failures of the Comet airliner. De Havilland and Walker, after reviewing this case comment:

'There is a modern trend which is steadily changing the overall character of investigations, though without affecting the basic principles. Accidents on the whole are becoming less and less attributable to a single cause, more to a number of contributory factors. This is the result of the skill of the designers in anticipating trouble, but it means that when trouble does occur, it is inevitably complicated.'

They point out that it is therefore futile to argue that particular explanations of failures are improbable, for in many cases, these are the only explanations consistent with known facts, and they conclude: 'This trend can make nonsense of public inquiries and of law-suits where allocation of responsibility is attempted' [58].

Reviewing these discussions of failures by mechanical engineers over a twenty-five year period, it is apparent that there is considerable awareness of and concern about the limitations upon the design process which are faced by the individual designer, although there are only one or two instances where there is any overt indication that social factors beyond the control of the individual designer might have contributed to the failures [59]. There are, however, numerous instances which make clear the considerable reliance which is placed in the design process upon psychological factors. Many areas of design, especially the assessment of risk, cannot avoid a reliance upon human judgement. The natural sciences may radiate an impression of distilled rationality, but even if science were like this [60], in all of the practical and concrete applications of such a science, it would have to become a part of the engineering design process, and thus become subject to all of the restrictions and limitations outlined above.

Engineers themselves are, of course, not unaware of the limitations which they are likely to encounter as they tackle the processes of engineering design, and two aphorisms cited during the course of a recent discussion of the problems of failure may serve to illustrate this awareness [61]. The first counsels simply that it is wise to: 'Expect one failure per annum per £10 000', while the second, perhaps rather more cynically, declares that: 'The number of failures equals four times the number of design staff on the project plus three times the number of senior staff'!

There is another, rather more general issue which bears upon this problem of safe and effective design, and this is the manner in which new designers can be trained to avoid patterns of practices which are likely to lead to failure. It is often assumed that all of the relevant and usable information which individuals can bring to bear upon a task or problem can be brought out into the open, stated explicitly, codified and, if necessary, set out in written form. This assumption was challenged to some extent in the above discussion of the difficulties of specifying goals in systems, and in the use of such goals to assess failures, and it needs to be challenged again in considering the manner in which designers acquire their knowledge and understanding of those practices which are likely to lead to failure, for similar difficulties about making precise and overt statements will be found in this field also. Michael Polanyi, the natural scientist and philosopher, has argued persuasively that there is a strong 'tacit' element in much knowledge, whether scientific or otherwise [62]. Men know more than they can say, he argues, and many of their practices, their approaches to problems and their responses to events derive from knowledge which is possessed by the body as an organism, and which cannot readily be articulated in a conscious manner. This is seen most clearly in the case of skills which are thought about only when they are being learned, but which are banished from mind when the body knows what is required [63].

Much scientific and engineering knowledge is of this tacit, craft nature, being absorbed and transmitted in the course of procedures of craft training. Personal contact and interaction provide the most appropriate medium for the transmission of such craft knowledge [64] (and also, of course, for the perpetuation of tacit misunderstandings about the world). Direct personal training, in science, engineering and design, as in many other fields, seeks to avoid failure not only by the statement of precepts and rules, but also by offering opportunities for contact with and discussion about 'pitfalls' which may lie in wait for the unwary novitiate. Practice during training in encountering and handling pitfalls builds up tacit knowledge about how similar cases may be handled in subsequent real encounters. To supplement such training or to attempt to compensate for its absence, some collections of exemplary pitfalls have been made: the amusing collection of letters in the 'Honeywood File', for example, presents a series of blunders and errors which architects may learn to avoid without necessarily being conscious that they are being presented with a lesson [65], and compilations of engineering failures such as that by Feld already mentioned may serve a similar cautionary function. It should be remarked, however, that knowledge of the kind being discussed here, being tacit, is not necessarily correct, and where a traditional mode of approach comes to be applied in changed circumstances, such implicit assumptions about practice may be the most difficult to correct.

Engineering, applied science and design generally, then, are very much *human* activities, in which aspirations towards an ideal of rationality have to be tempered by the limitations both of human abilities and of the 'state of the art'. Approximations, 'incrementalism', leaps of faith and other elements are acknowledged to recur in these fields, especially when novel or pathbreaking developments are being pursued, and successful practitioners in these fields may make use of a degree of intuition or craft or tacit knowledge which is difficult to codify.

Such procedures are naturally not infallible, and an awareness of this fact is important if an understanding is to be gained of the contributions which engineering and other forms of design may make towards the preconditions of disaster. Of equal importance, however, is the realization that the above comments are not limited to engineering or applied science, narrowly conceived, but apply to any set of practices by which men may attempt to manipulate or modify the world around them, whether these practices are concerned with town-planning, agriculture, landscaping, weather modification, canal-building or coal extraction. Thus, many of the insights gained by engineers in reviewing their own occupational experiences with the recurrent problem of failures may be scrutinized and transferred to other fields to help to provide a means of understanding those failures which are labelled 'disasters' [66].

2.4. *Accident Studies and the Management of Safety*

The study of disasters merges with the study of accidents, although for an accident to be labelled a 'disaster', it will probably need to be an unusually large-scale accident, an unusually costly accident, an unusually public accident, an unusually unexpected accident, or have some combination of these properties. But in spite of this overlap, those concerned to examine and understand accidents have not paid much attention to disasters as such. Within the somewhat self-contained specialism of 'safety and accident studies', concerns prior to the mid-seventies have been with the study and prevention of individual small-scale accidents [67], although it is possible that recent legislation in Britain and the shift in official attitudes indicated by the establishment of the Health and Safety Commission and the Advisory Committee on Major Hazards may produce some changes in the existing patterns of interest in safety [68].

In the world of safety, the major issues have been those relating to the interests shared by psychologists, ergonomists, safety officers and the staff of the various emergency services, so that when disasters or unusually large-scale accidents are discussed, it is usually with a view to the design and preparation of contingency plans to cope with the after-effects, should such an accident happen. Psychologists and ergonomists interested in the prevention of accidents have concentrated, not upon those of sufficient scale to be regarded as 'disasters', but rather upon the numerically more important and methodologically more accessible cases of smaller accidents affecting individuals or relatively small groups of individuals, typically road accidents or accidents in industrial situations

[69]. Studies in this area have progressed by means of the technique developed by ergonomists for studying 'near-miss' accidents, where little or no damage occurred [70], by the development of highly sophisticated engineering models for the analysis of individual and aggregate accidents and their causes [71], and by the development of theoretical psychological models.

In producing such models, psychologists have not neglected the pre-accident phase [72]. Most of them identify such a phase, and are aware that this phase may be studied independently in order to try to identify combinations of circumstances which are likely to precede accidents. But their discussion of the pre-accident phase naturally reflects their interest in the psychology of the individual, and in the interrelationships of the small number of individuals who are directly involved in the accident. As a result, their assessments of the pre-accident phase tend to be based solely upon a consideration of the situation in which one or two individuals find themselves immediately before an accident, and as it has recently been remarked, there have been very few attempts to locate accidents in their total situation, or to see them in the context of the wider social structure within which they occur [73]. A complex and sophisticated model which has recently been developed by Lawrence in order to direct his study of gold-mining accidents [74] illustrates the kind of attention paid by psychologists to this pre-accident phase. This model, which could well be extended to cover the wider range of features that a sociological analysis of the pre-conditions of accidents would need to consider, is set out in Figure 2.2 [75].

Where the findings of such investigations identify factors such as working conditions, worker selection, or the absence of safety equipment as contributing to accident rates within given work settings, the questions of whether such factors can be changed and of how they can be changed become problems for the management of the organizations concerned. The involvement of management in the safety area, however, has also tended to reflect the 'individual accident-prone worker' approach developed and pursued by ergonomists and psychologists, or the 'apathetic worker' approach taken by the Robens Report on Health and Safety at Work [76]. With regard to larger-scale accidents, or to accidents which arise from more complex and extensive causes, there seems to have been an assumption only that the manager should be ready for the catastrophe that *will* occur. The manager may thus readily obtain advice about the forming of emergency committees, for example, and about the preparation of rescue and relief plans [77]. It is often difficult, however, to define clear criteria for action in the implementing of such

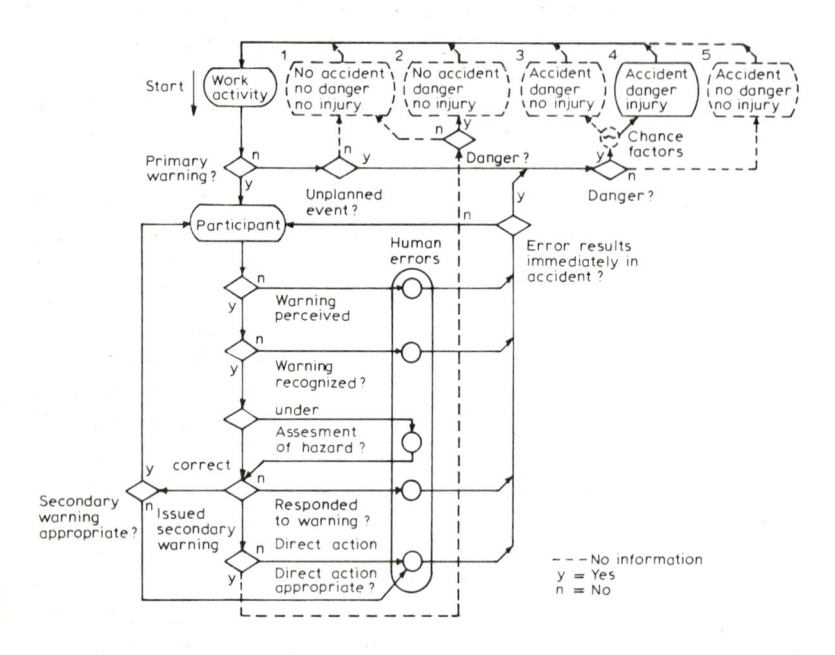

Fig. 2.2. Human error model developed to probe causes of accidents in gold mining. *Source:* A. C. Lawrence (1974), Human error as a cause of accidents in gold-mining, *Journal of Safety Research*, **6**, 78–88, Figure 2, p. 80.

plans, and there are a number of accounts which illustrate the difficulties of deciding when an emergency plan should be implemented, or of ascertaining whether a plan designed for one kind of emergency will necessarily be of use in another. City and hospital emergency plans were found to respond inadequately to the demands created by a tornado in Worcester, Massachussetts, for example, and by the notorious Cocoanut Grove nightclub fire in Boston [78], and considerable uncertainty was generated about whether a hospital emergency plan which envisaged many casualties needing surgery should be activated when the hospital in question was suddenly inundated with a large number of serious food-poisoning cases from a nearby conference centre [79].

In Britain, the need for management to devote much more time and energy to the consideration of such issues has been made clear by the passage of the Fire Precautions Act and of the Health and Safety at Work Act. Attention was focused much more dramatically upon such issues while this latter legislation was being drawn up, by the occurrence of a major explosion in a chemical plant at Flixborough, which prompted the formation of the governmental Advisory Committee on Major Hazards [80].

Within the decision-making settings in which managers operate, discussions of hazards and safety are regarded in terms of the risk from a known hazard, some estimate of an acceptable level of risk, and estimates of the costs of reducing risks to various levels [81]. To assist the manager, or at a higher level, the government policy-maker who is trying to take appropriate decisions about the amount of expenditure which can be justified in terms of the reduction of risk that will be produced, it may be helpful, particularly in such an emotive area, to try to quantify some of the issues, and tables of risks exist [82] which offer comparisons between the incidence of death which is likely, in an actuarial sense, from certain types of hazards. Convenient base lines for such tables are provided by estimates of the likelihood of being killed by lightning, or by radiation from the explosion of a distant supernova [83]. Such tables provide a means of focusing discussion upon known or expected risks, and upon the kinds of improvements which may be bought for a given outlay, although the notions of probability upon which such tables are based are not beyond question, especially when the hazards being discussed are new, or are infrequent in their occurrence. And, of course, where the incidence of a particular hazard is completely unexpected, it is unlikely to be considered in the deliberations arising from such tables of risk.

Some organizations with particular concerns about the hazards associated with their operations, and about the need to ensure public safety have taken an active initiative in using such methods of assessing risk to improve their operational levels of safety. The nuclear engineering industry and Imperial Chemical Industries are notable as organizations which have used such methods of assessing risks in order to change their operating procedures in ways which reduce risks from known hazards to acceptable levels [84]. At a policy-making level within industrial and other organizations, such wide-ranging discussions of the costs and benefits of one course of action as against another overlap with more general discussions of forecasting and risk in the more commercial sense of the word [85].

The responsibilities and concerns of managers are much different, of course, in those cases where their attempts to reduce or eliminate hazard fail, for they are then faced with more immediate problems of implementing their disaster plans and of maintaining and controlling the process of administration in crisis conditions [86]. Decisions then need to be taken about the plans which are to be activated, and the range of hazards which are likely to be faced must be assessed. Under such conditions, the deficiencies of disaster plans which have been constructed to meet hypothetical emergency situations become apparent, and the manager on the spot will expect to make rapid and immediate adjustments and corrections to compensate for earlier failures of forecasting. Some writers on management practices would see one of the central arts of management as being concerned with the ability to cope competently and coolly with such crisis situations, and it has been pointed out that it is possible to learn lessons for routine management from responses to such extreme situations [87]. Extending this argument further, organizational crises may, from one point of view, be considered as opportunities for pursuing social change within an organization [88], or even as means of aiding personal growth for individual managers [89]. The two Chinese characters used to express the word 'crisis' mean 'danger' and 'opportunity' [90]!

Such discussions are concerned more with the original meaning of the word 'crisis' as a situation in which important decisions have to be made in a particularly short time, than with disasters where management procedures must be maintained and management problems coped with under conditions of major technical emergency involving threats of injury and loss of life. It is true, though, that 'one man's disaster is another man's crisis' [91], and many professions specialize in handling routinely the crises or disasters of others [92]. In this perspective, it becomes natural to regard managerial crises as stages in the growth of a developing organization, and to include discussions of planning failures and of the failure of mergers in the same context [93], as part of a wider set of problems faced by organizations in a changing society [94].

With one or two notable exceptions, then, managers and other specialists concerned with safety have concentrated their attention upon small-scale or individual accidents and their prevention, and have tended to regard the development of emergency plans as an activity which is not closely connected with problems of tackling the understanding and forecasting of larger-scale accidents. The extended analyses of hazards which have been carried out by the nuclear engineering industry offer one way of developing a concern with the preconditions of disaster and these could be usefully linked with an extended form of Lawrence's psychological model. Beyond the field of hazard analysis, however, management writers appear to have been concerned with disasters more as occasions for testing and developing management skills than as occasions which should provoke a thorough analysis and review of methods of forecasting and decision-making within the organizations concerned.

2.5. Summary

In order to help us to begin to gain some understanding of the nature of disasters, we have reviewed the way in which a number of groups of specialists have approached disasters and related phenomena. We did not expect to gain much of direct relevance to an understanding of the nature and origins of disasters from accounts written by victims or survivors, and although some discussions of disaster by investigative journalists may provide valuable collections of data, such writers seem to be professionally inclined towards the particular, and away from the search for general regularities and principles which can be used to explain classes of events, rather than just single instances. Medical interest in disaster and large-scale accidents has concentrated upon the very apparent and very pressing problems of dealing with the range of injuries and public health hazards produced by various kinds of disasters, but recently one or two criticisms of this preoccupation have begun to be voiced by epidemiologists who wish to shift attention to an examination of the causes or origins of disastrous or disruptive events.

Managers and industrial psychologists have become accustomed to some discussion of accidents and safety, in the course of which they deal with the kinds of events which predispose particular individuals to become involved with particular kinds of accidents. In general, however, they have tended to adopt an individualistic approach to accidents, paying little attention to features of the wider social setting which also contribute to the occurrence of accidents, and it is clear that this individual-centred approach would need to be broadened in order to be used to examine the nature of disasters.

Of those specialisms reviewed in this chapter, it is the engineer, with his central and habitual preoccupation with failures and their origins, who seems most likely to offer a mode of approach which is helpful in starting to examine the preconditions of disaster. But while the engineer has a vocabulary for the discussion of failures, and a long-standing commitment to the search for practices and procedures which will reduce the number of failures in any given setting, his interest is naturally predominantly a technical one, so that he pays attention to social and administrative factors related to failures and accidents only when this is unavoidable. But the study of the nature and origins of disaster seems to be the kind of inquiry which might well be pursued by multi-disciplinary research teams, in which psychologist, epidemiologist, engineer and manager might well co-operate with social scientists, in order to understand the interrelationships between different kinds and levels of event which often seem to be instrumental in the development of disasters.

The area has seldom been identified as an independent area of study, so that few terms or ideas exist which are of direct relevance to the problem being considered [95]. One of the main aims of this book is to examine some sets of data relating to disasters in order to try to provide terms and theories which will serve as a kind of 'conceptual tool-kit' for the examination of disasters. Before this task is embarked upon, however, the largely technical approach to accidents and disasters which has been considered so far will be supplemented in the following chapter by an examination of the contributions which social scientists have made to the study of disasters.

3. Attempts to understand disasters—II

In the previous chapter, disasters were reviewed either as events which produced upheavals and injuries demanding medical attention, or as technical failures or breakdowns of an unusually large-scale kind. Neither of these rather narrow views copes fully with the range of events that we are concerned with when we come to look at disasters from a slightly broader perspective; for disasters arise from specific social settings, they are brought about by social and administrative incidents as well as technical ones, and the impact which disasters have upon society when they occur is not limited to the damage and injuries which demand medical attention. Disasters can only be fully understood if they are placed in the context of the social setting from which they emerge, and upon which, in their turn, they have an effect. For this reason, it is particularly important to include a consideration of sociological factors relating to disasters in our attempt to gain an understanding of these disruptive phenomena, and the present chapter reviews the work of sociologists in this area.

An extensive literature of social science disaster studies has grown up since World War 2, following the lead given earlier in the century by writers such as Prince, Queen and Mann, and Carr, who carried out studies of disasters in the 1920s and 1930s [1]. But while much useful material has been gathered as a result of these studies, it will become evident that, for anyone concerned to gain an understanding of disasters, and especially of the conditions which tend to be associated with the pre-disaster situation, there is a rather curious bias in the body of existing studies. Possibly this is because all of these disaster studies have tended to take an interest in the processes by which the 'disrupted social fabric' is repaired after a disaster; and possibly, too, as has been suggested recently [2], because disaster researchers are preoccupied with a 'bolt-from-the-blue' hypothesis about the emergence of disasters. Thus almost all of the studies take the *onset* of the disaster as the starting point for their research, treating and prior events in a cursory manner, if at all. This gives rather a lopsided look to the findings and, amongst the studies available, very few indeed offer any degree of assistance to the study of the nature, origins and pre-conditions of disaster, rather than the outcomes of disaster.

It was suggested above that the pattern for all of these disaster studies was set in the 1920s and 1930s, by social scientists who saw disasters primarily as examples of social pathology [3]. The onset of a disaster offered them an opportunity either to study a portion of the social structure in the abnormal and distorted form which it adopted under conditions of severe crisis, or to examine and chart the patterns of restorative social change which seemed to follow a major disruption.

One of the earliest studies is concerned with the devastation of the harbour and city of Halifax, Nova Scotia, by the explosion of a munitions ship in the harbour in 1915, the largest 'man-made' explosion which had occurred anywhere in the world to that date [4]. It destroyed much of the city, bringing an abrupt halt to commercial and political life, killing 2000 people and injuring a further 6000. Samuel Henry Prince, the sociologist who carried out this study, describes the consequences of the explosion, and then devotes his energies to detailing the manner in which, and the rate at which, the community returned to its previous level of functioning. He notes when the local newspaper resumed publication, when the railway began to operate, when the city council next met after the disaster, and how rapidly industry in the devastated city began to resume operations. Prince concludes by suggesting that the disaster did not have the effect of retarding the growth of the city. He takes the size and prosperity of Halifax five years later as an indication that the remaining members of the city had been stimulated by the destructive events to restore their community to a level which more than compensated for the losses which had been endured [5].

However, it should be noted that this story, with the relatively happy ending to which it is eventually pursued, begins for Prince with the moment at which the empty Belgian ship, the *Imo*, rounds a point in Halifax harbour and is confronted by the munitions ship, the *Mont Blanc*, steaming in the opposite direction. There is no mention of the reasons for the collision, nor of how it came about that a munitions ship should be carrying 450 000 lb (204 000 kg) of TNT, 35 tons of benzol and 2300 tons of picric acid packed in a manner so vulnerable to the danger of collision. There is no suggestion in the account that the origins of the event might form a part of the problem to be examined by a social scientist; indeed, Prince's approach tends to emphasize the completely unforeseeable nature of the incident:

'All were intent on the mighty task of the hour ... war gates closed, searchlights, forts fully manned, the gunmen ready.... The people knew these things; and no one dreamed of danger, save to loved ones far away. Secure in her own defences, the city lay unafraid and almost apathetic' [6].

Twelve years after Prince's study was published, Carr, in 1932, made another notable addition to the scientific study of disasters, by suggesting that it would be useful to distinguish between types of disaster according to whether their onset was 'instantaneous' or 'progressive', and according to whether their areas of impact were 'focalised' or 'diffused', distinctions which are still in use today as basic criteria for the classification of disasters.

But what is important here, in considering the relevance of Carr's work to a search for the factors which are associated with the origins of disasters, is the realization that Carr shares with Prince a concern for detailing only the nature and scale of the physical event and the physical and social disruption which it causes [7]. This concern makes it possible to examine the manner in which disaster is faced, to look at patterns of rescue and recovery and to chart the gradual return to normality, but it also takes it for granted that questions about the sources of the disruption cannot usefully be raised. There is rather a neat fit, or 'resonance' between this approach and the theoretical orientation of the writers concerned. For them, and for many other later students of disaster, the prevailing theoretical assumptions suggested that the interesting issues surrounding disaster were questions about how rents in the social fabric could be repaired. Existing functionalist theories could more readily account for the manner in which societal wounds might eventually be healed than they could cope with the problem of how society could come to wound itself in such a way in the first place.

The main body of disaster studies, which absorbed some of these concerns and assumptions, was produced after World War 2 [8]. Just as there was a medical interest in collecting together, and accounting for, the new sets of information provided by the exigencies of wartime activities [9], so social scientists seemed to be stimulated by the possibility that lessons could be learned from the disruptions produced by wartime events, and by the extremes of experiences encountered during the war period. Accounts of these events and experiences were available as data to be studied, and their ordering and examination offered one possible way in which the excesses which they reflected could come to be absorbed in an orderly and acceptable form into the wider culture.

There was a widespread awareness of the impact of the atom bombs at Hiroshima and Nagasaki as providing examples of 'ultimate horror' to be charted and recounted and, somehow, brought to terms [10]; but nonnuclear warfare and wartime imprisonment could also provide their own examples of 'ultimate horror' in the behaviour of human individuals and groups in extreme conditions; these, too, contributed to the post-war writings [11].

But the collective academic working through of the experiences of the war was not the major impetus for the main thrust of disaster studies in this period, as a consideration of one study of the period will show. Killian's study of the response to the explosion of a fireworks factory near the centre of Houston, Texas, in June 1953 [12] was one of a series of official studies carried out in America in the period 1945–63. The factory where the explosion occurred was located in a heavily populated area, about three miles from the centre of the business district of Houston, and it disintegrated completely in the explosion. There was heavy blast damage to buildings within a quarter of a mile of the factory, windows were shattered two miles away, and the explosion could be heard over a radius of 15 miles. Surprisingly for a major explosion on an urban site, only four people were killed, in a house near the factory, and injuries were also fairly light, most of the inhabitants of the neighbourhood being out at work.

As with other disaster studies, the report pays no attention at all to trying to explain how two workers in the fireworks factory inadvertently came to set off a firework; to how such an accident could spread to 80 000 lb (36 000 kg) of 'black powder' stored nearby; to how this amount of explosives could have been stored in two unmarked corrugated steel buildings near to houses in a densely populated urban area; or to how the city of Houston could come to have a city ordinance prohibiting the use of fireworks within the city limits, but no ordinance prohibiting their manufacture within the same area.

Instead, the report concentrates upon the curious fact that the explosion produced a mushroom-shaped cloud which could be seen from many parts of the city. Few people seemed to know of the existence of the fireworks factory; this fact, combined with the unusual appearance of the cloud, led some of the local people to think for a while that there had been a nuclear attack, although the extent of this response was exaggerated by the local press [13].

The outstanding feature of this disaster for the researchers was the 'highly unstructured, ambiguous situation' which existed immediately after the explosion, and their study was intended to discover whether the response of the inhabitants of the city to the explosion offered any clues as to their likely behaviour in the event of a genuine nuclear attack.

This short report of a rather unusual event exposes very clearly one of the assumptions which permeated disaster research in the post-war period. In the Cold War period, when the threat of a nuclear attack was seen as very real and immediate, the disruption caused by disasters offered an opportunity to try to explore some of the problems that might be created by a nuclear attack [14]. And these parallels could only be examined in the post-impact situation, so that there was no impetus to examine the origins of the disasters studied. If there was a problem about the origins of a nuclear attack, it was a problem for the diplomat and the political scientist concerned with the antecedents of a deliberate act of war. Thus the preoccupations of military defence reinforced the more general perceptions of disasters as essentially unforeseeable, and encouraged the view that the problems of disasters were located only *after* the moment of impact.

The set of disaster studies from which Killian's account has been taken provides information about a number of aspects of disasters: about warnings and how the warning messages can be effectively phrased and disseminated [15]; about attitudes to precautions such as fallout shelters [16]; about the functions of rumours and myths in disaster situations [17]; and about all forms of collective reactions to threats of disaster [18]. In addition to providing a response to the national concern with the threat of nuclear attack, such studies had more immediate benefits in that their findings could also be related to the problems of coping with the kinds of natural disaster prevalent in the United States, particularly to tornadoes, earthquakes and floods [19].

A considerable body of knowledge has thus been built up about how to inform and warn populations of major hazards which may threaten them, and of the action that they should take; about the likely effects of the impact of disaster upon various groups; about the immediate reactions of individuals caught up in extreme situations; about the manner in which spontaneous organization is likely to arise to deal with disaster; about the working of more formal relief organizations under disaster conditions [20]; and about the 'convergence effect' which leads people to crowd to the scenes of disaster [21]. Prince's original concern to examine the manner and rate at which the social structure and functioning of a community are restored after a disruption has also been pursued in many of these studies.

Most of this knowledge has been built up within an exclusively American context, and while there have been occasional cross-cultural studies, or studies located outside the United States of America, these external sources of information are often used principally to point up features of the American situation by contrast [22]. In one such study, of the reactions of the islanders of Yap, in the Western Caroline Islands, to typhoons, Schneider makes the rather piquant comment: 'In America, it is really difficult to see which has the greater impact, the tornado or the clean-up and rescue operations'[23].

Since the concerns of disaster researchers in the period following World War 2 were, like those of earlier researchers, with the impact and post-disaster phase, and since the problem of dealing with the effects of sudden nuclear attack mounted after a very brief warning period was also a related concern, there was very little demand at all for a theory to explain the causes or origins of the disruptive events studied.

A recent critic of this gap or omission in existing research has speculated on the possibility that society may 'need' disasters to enable it to 'vent internal pressures and transform anxious dread into a mass target-phobia' [24]. The degree of the neglect of the study of preconditions in disaster research certainly makes it necessary to give consideration to such speculations, and the manner in which the media deal with accidents and disasters could be taken as offering some support for them [25]. Western [26], in his discussion of the epidemiology of disasters, argues that a large body of information about the preconditions of disaster has not been built up, partly because of poor information collection arrangements, partly because of over-specialization amongst the various disciplines studying different features of disasters, but also because of the prevalence of a belief that all disasters are different. If those trying to examine and understand disasters are really trying to study a group of phenomena which have no common characteristics, at least in their early phases, they are clearly in a situation where science cannot operate, for no generalizations are possible, and they are dealing with a unique and unprecedented class of phenomena whose properties can never be predicted from past experience. However, it seems highly doubtful that this is, in fact, the case.

As Western points out [27], the aphorism 'each disaster is different' was coined by the League of Red Cross Societies to warn against stereotyped responses to the problems of the organization of relief, but it has been interpreted to mean that 'meaningful inductive conclusions are not possible'. In advocating the future application of epidemiological modes of analysis to the study of disasters, he very reasonably seeks to reject this interpretation, for it places an examination of the preconditions of disaster beyond the realm of scientific inquiry.

In spite of recent criticisms [28], reviews of the field in the 1960s [29] and in the 1970s [30] have remained resolutely wedded to a post-disaster orientation, maintaining a concern with reactions to disaster, and with rescue, relief and recovery, an orientation which allows no place for the study of the causes or origins of disasters.

In retrospect, it seems rather strange that there could have been such an extensive, and yet largely tacit, agreement that there was no point in devoting time and resources to the examination of the factors which led to the production of disasters, but that energies should be directed towards the problems of coping, and of mopping up afterwards. However, quite apart from the immediate and pressing problems created by major disasters and disruptions, particularly where there is extensive injury and loss of life, the reasons for the persistence of the post-disaster emphasis may become more understandable if some of the theoretical models developed and used by disaster researchers are examined, for these models enshrine the built-in assumptions which researchers in this area would not normally need to challenge.

Thus for example T. E. Drabek [31], tackling the general issue of the kinds of methods appropriate for the study of disasters, proposes to classify all methods within a three-dimensional space, which he represents in a diagram similar to that shown in Figure 3.1. The three

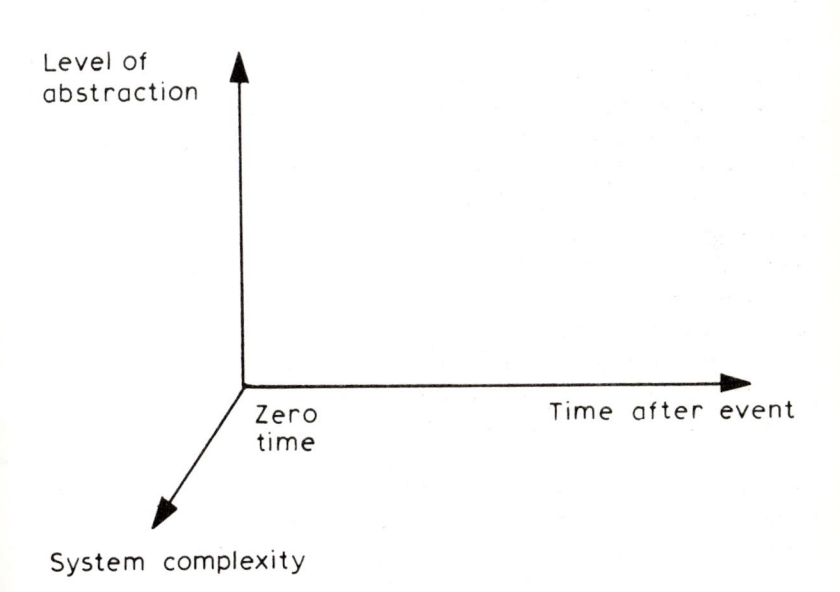

Fig. 3.1. Schematic representation of Drabek's theoretical framework for studying disasters. Adapted from T. E. Drabek (1970), Methodology of studying disasters: past patterns and future possibilities, *American Behavioral Scientist*, **13** (No. 3), 331–343.

dimensions of his classification-space are 'system complexity', 'level of abstraction' and 'time after event'. Clearly, since his 'time after event' scale starts at 'zero' at the moment of impact, there is no conceptual room in his scheme for any methods which might be required to look at the pre-conditions of impact.

This is not to say that disaster researchers have assumed there is no time period relevant to their studies which might be located before the moment of impact; but when earlier time periods are considered, they are all interpreted in the light of the assumption illustrated so clearly by Drabek's diagram: that the 'real' time period worthy of study, the period where the substantive problems are to be found, is after the moment of impact.

The influential and much-quoted models proposed by Powell and his colleagues [32] do pay some attention to the period before the moment of impact, but, as Figure 3.2 makes clear, they admit only the possibility of 'warning' and 'threat' phenomena in this period. Since these phenomena are of interest only insofar as they concern the behaviour and reactions of possible *victims* of the disaster in the period immediately before they are affected, they again reflect an interest in the problems of reaction and impact, and do not serve to direct attention to any possible causal factors.

(a) without warning

| Impact | Inventory | Rescue | Remedy |

Time →

(b) with warning and threat *   Powell *et al* (1953)

| Warning | Threat | Impact | Inventory | Rescue | Remedy |

* Note: Powell *et al* present other variants, with the same post-disaster emphasis

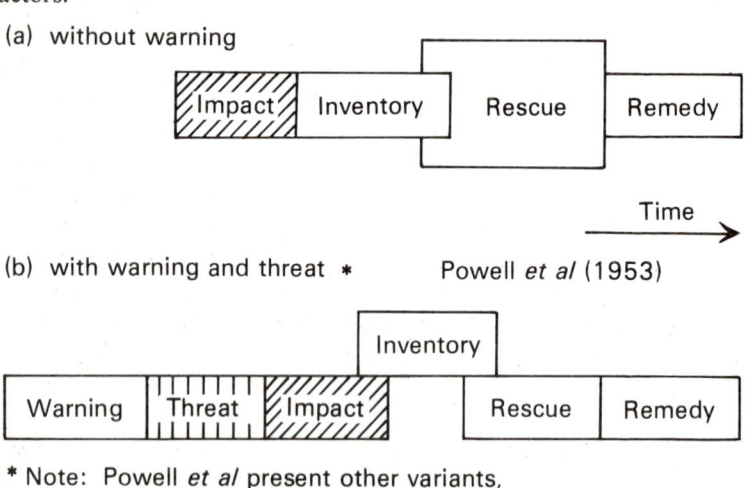

Fig. 3.2. Schematic representation of the theoretical framework of Powell and his colleagues for the events associated with disasters. Adapted from J. W. Powell, J. Rayner and J. E. Finesinger 'Responses to disaster in American cultural groups' (see note 32).

In a similar way, when Barton, in one of the most comprehensive and authoritative statements in the post-war disaster research literature, proposes that he will not deal with the pre-disaster period because of lack of material, his statement of omission, too, reflects the concern with the post-disaster period which is expressed in his examination of the restorative processes of the social systems which emerge after disasters. His statement about the pre-disaster period reads as follows:

'The entire question of pre-disaster preparedness, on which there is relatively little research, will be left out, as will the subject of the detection of threat, the issuance and transmission of warnings and the response to warnings, on which there is a great deal of research'[33]. The assumption is clearly expressed that the only phenomena worthy of note in this period before the disaster are those related to the 'pre-impact condition of the victims'.

The assumptions so far discussed show the manner in which considerations of the origins of disasters have been left out of the chronicles of disaster, and if attention is shifted from time to space, the maps of disaster are also found to display the same bias, for those studies of disasters which set out spatial models for the study of disasters typically present a picture of concentric annular zones spreading out from the point of impact, whether these zones are conceived of in epidemiological terms, or in sociological terms.

Thus Anthony Wallace [34] presents, in circles which should be taken as representing a conceptual rather than an accurate physical or geographical distribution, a zone of total impact at the centre of his 'map' of disaster, surrounded by a zone of fringe impact, beyond which is a zone of filtration, to which the injured and refugees move. Further away from the area of immediate impact, Wallace delineates a zone of organized community aid, where the formal organization of rescue and relief operations is set into motion, and beyond this, he sees a further zone of national or international aid. Taylor and Knowlenden present an essentially similar model within an epidemiological context [35].

3.1. *Threats and Warnings*

Social scientists interested in disaster, then, have tended to concern themselves with the major and pressing questions of how disasters make their effects felt and with the problems of counteracting or ameliorating these effects. Given this concentration of concern, it is difficult to find any helpful information at all in the social science literature which relates directly to the problem of the origins of disaster. Some individual accounts of particular events do discuss the causal patterns responsible for the incident in question, but with very little awareness of the possibility of developing these comments into general statements [36]. Perhaps the only writer who has tackled this problem seriously is Wilensky [37], who examines, not an inadvertent disaster, but the failure of the American intelligence and military services to analyse and collate the information available to them about the impending Japanese attack upon Pearl Harbor, using as data the detailed accounts of events provided by Roberta Wohlstetter [38]. But setting aside this political science study, the only remaining social science writings in the area are those concerned with warnings and threats.

For many individuals and groups, an encounter with an unfamiliar, unknown situation is likely to generate anxiety [39], and the typical response to threats or to warnings of danger may be considered to be extensions of, or reactions to this kind of anxiety [40]. The anxiety is less, of course, if the threat is a recurrent and familiar one, and individuals react less anxiously to threats which can be contained in one sector of their life-situation, without spilling over to threaten other concerns. Also, when the implications of a threat of danger can be readily discerned and predicted, the response to the threat is likely to be less extreme. The maximum degree of anxiety is likely to be provoked by threats which relate to an unprecedented danger, whose effects appear to be unpredictable, and whose impact seems likely to be distributed in an uncertain and undiscriminating manner [41]. Fortunately, threats of this latter kind are rather rare.

Another way of considering the threat of impending danger is to regard the threat as a piece of information, or as a set of several pieces of information which must be absorbed and processed appropriately by a psychological or a social system [42], if an effective response to the threat is to be considered and developed. When the information input process is not functioning as effectively as it might, because of an excessive amount of incoming information [43], or because of distortions in the transmission or reception of information, the response to the threat is likely to be inadequate. Groups under threat develop their collective response partly in the light of the manner in which news of the threat, or of anxiety about the threat is spread through the group. Such communications between individuals in a threatened community are not, of course, restricted to the transmission of information about the danger which is perceived; they also serve to spread particular types of evaluative and emotional response to the impending danger, to develop a unified (and sometimes distorted) group perception of the nature of the threat, and to transmit judgements about the actions or responses which various members of the group feel are called for [44].

Where the groups which are threatened are small, and dispersed, with not very much internal solidarity among members, the disruption produced by a major threat is likely to be considerable, although even with larger, more concentrated and cohesive groups, certain types of threat may still provoke considerable disruption of the existing social arrangements. The most disturbing kinds of threat are those which relate to rapid and unfamiliar destructive forces which strike without warning, and the degree of disturbance is likely to be enhanced if there is much destruction, and if the condition persists for a long time [45].

In extreme conditions of this kind, individuals and groups appear to find at least a minimal degree of security by interpreting the new situation in terms of a single dominant explanation, usually one which draws upon familiar concepts, arguments or ideas about the world. Of course, this tendency to seize upon a stable, familiar and dominant explanation of the shattering and unprecedented events which affect groups in disasters may lead to their adopting interpretations which seem inappropriate to outsiders who come in to offer help, and their actions may similarly seem inappropriate to those not directly involved. The suggestibility of individuals and groups in such stressful situations is high, and prolonged subjection to puzzling and incompatible cues about what is happening to the once-familiar surroundings tends to produce depression. Sorting out and resolving this conflict can produce an elation after the depression, as a stable and durable understanding of the world and the individual's situation in it is regained, and one of the most effective ways of producing this kind of outcome in a new situation is for the individual concerned to try to do things, to start to act and intervene in the new situation in which he finds himself [46]. Even then, particularly where the new situation is a technically complex one, there remains the possibility that these exploratory actions may themselves be hazardous in some way, and in situations of danger, in many areas of modern society, even calm, intelligent and reasonably well-informed men may not be able to discover what action is appropriate to avert or alleviate further damage [47].

But the problem of dealing with the anxieties provoked by menacing danger and of coping with the stresses of danger realized is not merely a problem of sorting out accurate perceptions of what is happening, and calculating the correct action that the situation calls for. Both the threat of danger and the consequences which follow when disaster strikes, provoke responses in individuals which involve their whole being, for the total organism is threatened, and the total organism responds. Martha Wolfenstein, in an absorbing discussion of disaster from a psychoanalytically-oriented point of view, has explored and discussed many such responses, using material from studies of tornadoes, floods and explosions, as well as wartime evidence relating to the response of civilian populations to both conventional bombing and nuclear attack [48].

Most individuals suffer little anxiety, she notes, when they consider that the possibility of danger is remote, and they ignore warnings under such conditions. They may have a variety of reasons for not heeding warnings of distant danger. One of these, Wolfenstein suggests, is the sense of personal invulnerability which it is essential for most individuals to maintain if they are to be able to go about their daily business without constantly worrying about all of the possible dangers that could threaten them. Also, even when danger comes a little nearer, the acknowledgement of a threat as real may involve inconvenience, especially if it then means that possible evasive or precautionary measures need to be reviewed. An individual concerned about his home and property, for example, may resist acknowledging the significance of a threat which could mean that he would have to abandon his house. When a danger is remote, a normal individual is most likely to respond to threats or warnings by denying the danger. But as danger comes nearer, and as the pressure to recognize the danger and acknowledge it by some kind of precautionary action becomes more acute, denial becomes more of a pathological response.

The ways in which people respond to warnings of danger and instructions about precautions issued by authorities are complicated, too, by the fact that these messages may evoke responses which were originally developed as responses to parental figures. One example of such a reaction is provided by the observation that windows at the fronts of houses in London during the blitz of World War 2 were 'blacked out' more fully than the windows at the backs of the houses. The front windows could be seen from the street by the official wardens, but, of course, enemy bomber pilots could see light from either side of the houses! Many other examples cited by Wolfenstein illustrate clearly the propitiatory element which is often associated with the taking of precautions. Those who take precautions often seem to have the feeling that they should be safe because they have obeyed instructions, because they have been 'good', rather than because they have taken the action which an objective assessment of the risks indicated. And, by the same token, if they are harmed, they are likely to feel anger and resentment against the authorities for not protecting them adequately, especially if they have complied with instructions.

Wolfenstein draws attention to the way in which individuals who are affected by disaster tend to perceive events as though the events were directed solely at them—'the illusion of centrality'. Regardless of the actual chances of injury, those who undergo the experience of standing close to danger, and of feeling it directed at them are likely to revise their evaluation of this particular hazard. Whether the individual in objective terms was likely to have been harmed or not, the crucial psychological distinction which Wolfenstein makes relates to the subjective experience which occurs. If the individual himself is convinced that he has been in a 'near-miss' situation, his subsequent response to the danger is likely to be changed.

After a disaster, those survivors most directly affected have been observed to show a characteristic 'disaster syndrome' of behaviour. They are unresponsive and do not display emotional reactions. Their activity is inhibited, and they are docile and undemanding, with an apparent resistance to taking in any more. As these symptoms disappear, their reactions may vary, Wolfenstein suggests, according to a variety of factors, including cultural factors, the survivors' sense of loss with regard to others who may have perished, and their own relief and guilt at being spared. Indeed, the extent to which they are aware that others have been affected, the extent to which they regard their suffering and dislocation as their own personal burden rather than as part of an affliction experienced by a wider group may have a significant effect upon their behaviour.

Some of the factors mentioned by Wolfenstein show how the process of giving warnings about impending danger may be complicated and, at times, made ineffectual. However, because of the calls for practical advice to be given to those responsible for issuing warnings, and particularly, as was indicated earlier, because of concern since the war with the problems of warning populations of the action to be taken in case of a nuclear attack, a number of researchers have examined the problem of giving adequate warnings, and of overcoming some of the difficulties mentioned by Wolfenstein.

In general, the success of a warning depends first upon it reaching those for whom it is intended, and secondly upon their reacting to it in an appropriate manner. In turn, the response of the recipients depends upon their present situation and past experiences, and this response will be influenced by the manner in which the warning information is detected, interpreted and passed on to them by others.

A warning must be *initiated*, must be *coded* into an appropriate form, and then *transmitted*. Those issuing the warning should then be concerned to assess the success of their warning by examining the *feedback* which they receive, and by looking at the *responses* of those who have been warned [49]. Difficulties may arise at any of these stages; in particular, there may be considerable uncertainty surrounding the decision to issue a warning, for it is often difficult to settle upon desirable criteria which indicate whether a warning is needed, and to discern when these criteria have been met. There is an interesting journalistic account of the uncertainties experienced by the staff of the Boston Weather Bureau in trying to assess whether or not a tornado warning should be issued [50], and a more extensive study of 'tsunami' warnings reinforces this point. In this study [51], W. A. Anderson examined the responses of two communities to warnings of danger from 'tsunamis', powerful seismically induced ocean waves which increase in size as they approach coastal waters, to constitute a considerable hazard to communities along certain coasts. In Anderson's communities, repeated warnings of disaster which did not materialize led to a lack of public willingness to co-operate in the actions which subsequent warnings called for, and Anderson concludes that there is a need for clear criteria to be established about the adequacy of the information needed before a decision to issue a warning is taken.

There may be difficulties, then, in deciding precisely when it is desirable to issue a warning, for if an error is made, misleading information will be transmitted. Warnings may also be misleading if the information they contain is modified by delays or distortions in transmission. The public for whom the warnings are intended may have some awareness of the possible errors which can give rise to faulty warnings, particularly if they have heard 'Wolf!' cried too often before, and this awareness may mean that the warnings will be ignored. Messages may be misperceived, it may be assumed that the warning is meant for someone else, or the warning may arrive too late for effective action to be taken. When a warning has to be passed on by intermediaries, the understanding of the meaning and authenticity of the warning may be qualified by evaluations of the characteristics of the intermediary, their reliability, possible motives for presenting such information, and so on [52].

Even when the warning is transmitted accurately, the kinds of factors mentioned above make it likely that the recipients of the warning will tend to attribute it to familiar causes, to feel invulnerable and to trust the authorities to provide safety [53]. For these reasons, Withey, in looking at the problems of providing adequate and effective warnings, has suggested that the receipt of a warning is the starting point, rather than the end point, of a series of stages. A person receiving a warning, he suggests, will first need to verify the message, if there is time available, and then to authenticate it. Having satisfied himself that it is a genuine warning of real danger, he will then need to elaborate the message in order to determine the nature, type, severity, location, timing and probability of the danger and to clarify any additional information needed. The individual will need this extra information to enable him to determine whether the threat is manageable, or whether it can be postponed. Alternatively, he will want to assess whether there are means of escaping the danger, and what chances there are that it can be endured and survived. On the basis of this kind of information, it then becomes possible to work out an appropriate form of response to the danger [54]. Withey's model is somewhat speculative, and it presents the stages which need to occur, rather than those which occur in actual instances, but it provides a useful reminder of the complexity of the warning response process.

Because of these complexities, it is clear that warnings should be accurate, unambiguous and reliable [55]. However, it is also clear, as Williams concludes, that 'it is difficult to warn people successfully against impending danger when they have not experienced the predicted danger in recent years and when they cannot directly perceive the signs of danger' [56]. To counter these difficulties, he suggests that warnings should be made precise, specific and localized. Prior action is desirable to check that some consideration has been given to conceivable dangers, and that plans of action exist, and training and practice in the response to warnings should be carried out. When a warning is issued, it should be a 'call to action' rather than just a sign of imminent danger, and when there is a threat of great danger, the need for confirmation of the message should be appreciated. That is to say, in such circumstances, it is desirable to present a 'This is it!' message, and to follow with a 'Yes, this really *is* it!' message. The process of warning police and personnel in other emergency services also needs to follow the same pattern [56].

The problems of warning and threat in the pre-disaster period, then, have received some consideration from social scientists, and a number of extremely useful findings have emerged from work in this area. However, it is clear, as we noted earlier, that all such studies concentrate upon the awareness and the activities of the victims or potential victims in the period before a disaster; and since, in most forms of disaster or large-scale accident, the victims are not responsible for causing the accident—or if they are, they only provide the last link in a chain of contributing events—it is evident that such studies will not add much to our understanding of the manner in which disasters come about, even though they do pay some attention to the pre-disaster period. There seems, therefore, to be a need to pay attention not only to the technical factors which are associated with the failures leading to disaster, but also to try to combine this concern with an examination of the social factors which are at work at the same time.

3.2. *Summary*

Attempts to gain an understanding of the processes, and particularly the social processes which are associated with disasters have been made for more than fifty years, but the focus of such studies, virtually without exception, has been upon the *impact* of the disaster, and upon the problems of rescue, relief and recovery. When the immediate problems of the aftermath have been coped with, the social scientists have switched their attention to the manner in which the community or group affected by the misfortune attempt to restore their condition to its previous level. Where studies have not followed this emphasis upon the aftermath of disaster, and have looked at events prior to the disaster, it has been in a context where 'disaster' has tended to be equated with 'surprise nuclear attack', so that the problems which have been examined here have been those relating to the provision of clear and effective warnings.

The range of studies reviewed in Chapters 2 and 3 indicate that although information is available about many aspects of disasters, much of this data has been collected by investigators whose main interests have not been centred upon the understanding of the development of disasters as such. Medical and sociological writers have tended to concentrate upon different aspects of the aftermath of disaster, and whilst engineers are more interested in the pre-conditions of failure, this interest is normally found to be limited to the technical pre-conditions of technical failures. If disasters are to be understood as developing socio-technical events, which do not arrive as 'bolts from the blue', but which are amenable, in many cases, to reasoned investigation and explanation in terms of generally applicable relationships and regularities, it is clear that there is a particular need to look at the combinations of social and technical circumstances which are associated with disasters and large-scale accidents, and to look especially at those circumstances which precede and contribute to the development of disasters.

4. Three disasters analysed

Some reasoned consideration of the kinds and combinations of circumstances which are likely to give rise to disasters seems to be urgently required at the present time, but there seems to be only the slightest awareness that this area is one which could be regarded as a unified area for study, or that research in this area could give rise to useful general principles in the study of disasters. So little work has been carried out in this field that any study at the present stage can only be an exploratory one.

In this chapter and in Chapters 5 and 6, some sets of data drawn from British Government sources published between 1965 and 1975 will be examined in order to try to discover what kinds of phenomena may be associated with the pre-conditions of disasters and large-scale accidents, to label and order these phenomena, and to test out the insights developed against new sets of data. In the process, a theoretical framework is developed which, although conceived in sociological terms, should be sufficiently detailed and sufficiently close to the empirical base from which it was drawn to make it useful to anyone interested in the field, whatever their background.

The primary intent at this stage is to develop a firm basis for a theoretical understanding of this novel area of investigation, and the pursuit of evidence had this end in mind, rather than the aim of producing comments about the incidence or distribution of different kinds of pre-disaster conditions. These latter issues can be more readily pursued when a better understanding of the basic processes is available.

4.1. *Three Examples of Disaster* [1]

The initial step in the investigation was to carry out a detailed examination of the pre-conditions associated with three major disasters. The first consideration which led to the selection of the three incidents chosen as starting points was that a detailed published account was available in the form of an extensive report of a public inquiry for each incident. But this was also true for many other incidents, and a little more could perhaps be said about the reasons for this initial selection.

In one sense, in an exploratory analysis, the rules for selecting material for analysis are not very strong, since these rules can only be developed after some degree of understanding of the field has been gained through analysis and observation. But there was one factor which seemed to offer some guidance. Work in a small number of industrial organizations had previously indicated that in the handling of many problems, particularly those which involved many groups and individuals, there existed a condition which has been called a 'variable disjunction of information' [2]. This term is discussed in more detail below, but it refers generally to a complex situation in which a number of parties handling a problem are unable to obtain precisely the same information about the problem, so that many differing interpretations of the situation exist.

It was felt that this kind of condition might well be found in pre-disaster situations, or situations where there was a 'failure of foresight' [3] developing as a result of the interaction of a number of groups and individuals.

For this reason, the three public inquiry accounts selected for initial analysis were consciously chosen because they related to accidents arising in conditions where a large, complex problem, the limits and bounds of which were difficult to specify, was being dealt with by a number of groups of individuals, usually operating in separate organizations, or in separate departments within organizations [4]. The value of the analysis presented lies in the extent to which it is judged to offer a satisfactory basis for handling discussions of incidents other than those on which it is based, and in the extent to which it advances understanding in later stages of the research [5].

This chapter will present, first of all, a more detailed account of the idea of 'disjunct information', and the main features of the first three disasters to be examined will then be outlined. The major observed similarities between these incidents will be set out in the form of a series of preliminary classifications or categorizations. As further evidence is discussed in subsequent chapters, these preliminary sets of characteristics will be grouped together into more useful patterns of relationships. The main concern at this stage, however, is to present an initial account of some of the basic kinds of phenomena likely to be encountered in studies of the origins of disasters.

4.2. *Variable Disjunction of Information*

In an earlier examination [6] of the behaviour of managers and others concerned with the scheduling of work through batch production factories, it was discovered that at the heart of this rather mundane and everyday technical task, there was quite a complex problem of handling and processing information, for the problem of finding the best way of scheduling batches of work through a typical small engineering factory was a problem with an infinite set of possible solutions. Since such problems cannot be solved directly, in most batch-production factories concerned with this issue the need to solve the problem is circumvented by the semi-tacit adoption of a series of rules-of-thumb which cope with the practical decisions that need to be made [7].

The significance of this for the present discussion is that this kind of information-handling problem did not seem to be one which would be encountered only in production scheduling, but that what had been studied was a particular example of a more general and more widely distributed condition which would be found among many groups. Within the scheduling departments of the factories studied, methods of coping with the unsolvable nature of the scheduling problem had had to be developed by trial and error over a long period of time, for the problem recurred every day, and had to be resolved somehow. But it seemed likely that in other situations, where potentially destructive forces were being handled, and where there was less opportunity to discover suitable modes of response by means of trial and error, such an information condition might well be found to be one of the features which preceded disaster, and this possibility was kept in mind in the selection of the first three cases for examination.

In the complex information-handling situation which is being discussed, several groups are involved, trying to manipulate a state of affairs for which they are unable to agree upon a single, authoritative description. Because each person has access to slightly different information, each one tends to construct slightly differing 'theories' about what is happening and what needs to be done. It is possible, of course, to reconcile all of the conflicting aspects of these theories to produce an agreed one, if sufficient time, money and energy is available, but it is typically the case in such situations that while complexity and uncertainty are considerable, time, money and energy are scarce.

The general condition is thus one in which the amount of information which can be generated or attended to with available resources is far less than the amount needed to describe or take account of the complexity of the situation in full. Relevant information, or perhaps what Wilensky calls 'high-quality intelligence' itself becomes a scarce resource in such circumstances, so that the cost of obtaining one piece of information has to be balanced against the lost opportunity of locating another piece of information. The condition is called a state of 'variable disjunction of information', therefore, since it is not a static state. Information is constantly being exchanged or amplified to remove discrepancies or to clarify ambiguities. But the central point to be realized is that the cost of information needed to reconcile information in one area, or between one set of parties, precludes the expenditure of resources in other directions. This condition cannot just be dismissed as a 'lack of communication', for it occurs in situations in which high complexity and continuous change make it necessary to be extremely selective in the use of 'communications'.

Where a number of groups or individuals are concerned to deal with well-structured problems [8] oriented towards specified goals and readily described in numerical terms, they are likely to be able to develop routines for the solution of these problems, if such routines are not already available. In such conditions, a variable disjunction of information is not likely to be found. But many problems in the modern world, particularly those surrounding the handling of forces which give rise to hazards, are much less well structured. Where problems use symbolic or verbal variables, have vague, non-quantifiable goals and lack available routines for their solution, relying instead on *ad hoc* procedures, a variable disjunction of information is more likely to be found. As the discussion in Chapter 6 will show, disasters may often be regarded as arising from attempts to handle ill-structured problems, the full implications of which were not realized before the event.

With the above notions in mind, the initial search for case studies of disaster proceeded by scanning reports of disaster inquiries over the past ten or fifteen years, and setting aside for later analysis those reports in which the official inquiry concluded fairly readily that the failure was attributable to some previously totally unknown factor, or to some clear lapse from good practice [9]. The effect of this search procedure has been to ensure that the cases initially examined were complex rather than simple ones, and while the kind of analysis developed clearly reflects this, the analysis is also one which makes it possible to include the less complex cases at a later stage as special instances. The empirical examination of data began with a detailed consideration of three cases of 'intelligence failure' in complex circumstances which promised some similarities with the condition of variable disjunction of information already described. These three cases were the report on the Aberfan disaster in Wales [10], the report on the Hixon level-crossing accident [11], and the report on the Summerland leisure centre fire [12].

4.3. The Three Disasters

In the first incident, a portion of a colliery tip on a mountainside at Aberfan slid down into the village, in 1966, engulfing the village school, and killing 144 people, including 116 children, 109 of whom were in the school. In the second incident, at Hixon in Staffordshire, in January 1968, a large road transporter, 148 feet (45 m) long, carrying a very heavy transformer, was hit by an express train while it was negotiating a new type of automatically controlled half-barrier railway crossing, killing three railwaymen and eight passengers on the train. The transporter moved at 3 feet/second and therefore could not clear the crossing in the 24-second warning period. In the third case, a holiday leisure complex at Douglas, Isle of Man, with approximately 3000 people inside, caught fire, on 2 August 1973. The building, an open structure clad partly in sheet steel and partly in acrylic sheeting, burned rapidly and 50 men, women and children in the building died [13].

The common feature which forms the initial starting point for discussion in the Aberfan, Hixon and Summerland disasters is that in each case the accident occurred when a large, complex problem, the limits and bounds of which were difficult to specify, was being dealt with by a number of groups of individuals usually operating in separate organizations, and separate departments within organizations. Thus, at Aberfan, we may suggest that the 'ill-structured problem' was the running of a pit and its ancillary activities to the satisfaction of H.M. Inspectors of Mines, of the various departments of the National Coal Board, the pit employees and local residents and their elected representatives. At Hixon, the problem was the introduction and operation of a new type of level crossing to the satisfaction of the various departments within British Rail, the Ministry of Transport, the police and the wide range of road users, including children, farmers and those likely to use the crossing with animals, agricultural machinery or other abnormal loads. Finally, at Summerland the problem tackled was one of building a new, attractive, profitable and safe form of lesure centre, using some new materials.

For each case, the dominant factors upon which the inquiry concentrated are summarized below, together with a brief indication of the organizational units and sub-units involved in each incident.

4.3.1. Aberfan.

The Tribunal of Inquiry considered the part played by a number of bodies and organizational units in its lengthy assessment of the events leading up to the tip slide. In addition to evidence taken from a number of technical advisory bodies, the Tribunal heard evidence relating to the organization of the colliery where the tip was sited, and to the role of the Area and Divisional bodies responsible for this colliery. Evidence concerning the involvement of the Headquarters of the National Coal Board, the activities of a committee set up to review organization within the Board and the part played by the National Union of Mineworkers was scrutinized. The Tribunal was particularly concerned to establish facts about the policies pursued with regard to the siting of tips, and about the manner in which information about earlier, non-fatal tip slides had been disseminated within this very large organization. On this last point, the Tribunal discovered that knowledge about the procedures necessary to stabilize tips had been available for many years. An engineer in a local company which was subsequently incorporated into the National Coal Board produced a memorandum in 1939 which anticipated the causes of the disaster. But subsequent circulation of this document within the National Coal Board was restricted to a small number of the professional engineering groups within the Board. This meant that the nature of the problem was not generally appreciated, in spite of the occurrence of other tip slides in post-war years, and it is notable that the London Headquarters of the National Coal Board remained unaware that tips constituted a potential source of serious danger until after the Aberfan incident.

Organizations outside the National Coal Board were involved when local residents protested, through the local Borough Council, about the possible danger from the tips at Aberfan. The local Planning Committee and the Borough Engineer's office were concerned in negotiations with the Board, assisted to some extent by the local Member of Parliament. The Tribunal considered at length the unsatisfactory manner in which these representations were handled locally by the Board, and their failure to reach a satisfactory outcome.

Finally, the Tribunal took much evidence from management and workers at the colliery at Aberfan about the perfunctory manner in which the decision to build a tip over a small stream had been taken, and about the response of management and workers to the various warning signs which began to be apparent as this tip grew in size and the slip became imminent.

However, although the situation was complex, and there were many contributing factors, the Tribunal of Inquiry found that the dominant pattern which contributed to the disaster was one which was located in the National Coal Board, and in the coal industry more generally, and which might be characterized as a pervasive institutional set of attitudes, beliefs and perceptions which led to a collective neglect of the problems of safety relating to tips by almost everyone concerned.

This collective neglect had a number of components: there were historic and institutional precedents in the neglect of tips by the 1938 Commission on Safety in Mines, and by H.M. Inspectors of Mines and Quarries; these were reinforced by the existence within the coal industry of sets of industrial beliefs which gave little consideration to tips; in consequence, the perception of potential dangers associated with tips was dimmed, even when slips occurred elsewhere. Few staff were appointed to deal with the problem of safety in tips, organizational practices were also oriented away from the problems of tips, and towards those of mines, and such literature as existed on tip safety was neglected or not given wide circulation. Naturally enough, this powerful bias in attitudes within the coal industry affected the patterns of decision-taking in relation to tips, so that when, for example, decisions about the siting of new tips were being made, little trouble was taken over this problem.

4.3.2. *Hixon.* The investigation into this incident drew upon evidence from a substantial number of organizations concerned. Within British Rail, evidence was taken from the train crew, from those departments responsible for planning and implementing the introduction of the new forms of automatic crossing, from those responsible for disseminating publicity to potential users of the crossings and from those who installed, inspected and modified the Hixon crossing. Within the Ministry of Transport, evidence was taken from the Railway Inspectorate responsible for approving the new crossings and the procedures associated with them. Within the Police Force, those responsible for circulating information about the new crossings, and those responsible for briefing police patrols escorting abnormal loads were questioned, as well as the policemen who were escorting the load which was in collision.

The communication links between the Ministry of Transport and the Police through the government department responsible for police affairs also came under scrutiny, as did communications between the police and a unit within British Rail responsible for bridge safety, which checked routes for abnormal loads with regard to possible dangers to weak bridges, but which was not explicitly required to consider such hazards as automatic level crossings. Evidence was also taken from members of two commercial organizations: the company which owned the transformer which was in transit, which had a factory adjacent to the automatic level crossing; and the haulage company which owned the transporter. Within the latter company, the patterns of responsibility and awareness of statutory obligations with regard to the transport of abnormal loads were examined, and the inquiry also considered communications between top management, middle management and the transporter crew, and communications between top management and British Rail over a previous incident concerning a lorry stalled on an automatic level crossing.

Within this complex set of organizational responsibilities and communications, the central and most distinctive contributing feature leading up to the disaster was a failure on the part of a large number of individuals in British Rail departments, in the management of the road hauliers concerned, and in the Ministry of Transport to bring together creatively the information which they all had, or had access to, in a way which would have made clear the danger of the new crossings to a long, slow-moving vehicle which was already part of the way across an automatic half-barrier crossing when it began to close. Many other precautions had been taken but not this one, and this was the factor which led to the accident. This failure, which we might, perhaps, characterize as a failure in creative problem-solving, was compounded by what we might call a 'passive administrative stance' adopted with regard to the new crossings by other parties involved, notably the police, who had received the information necessary to avert the disaster, but who had failed for a number of reasons to consider it actively. A further contributing factor was the behaviour of the drivers and crew of the transporter who could have acted to avert the disaster, had they been alerted to the need for this, and to the procedures called for. They, in a sense became involved in this tragedy as members of the public who were expected to take a responsibility for their own behaviour in relation to the hazard of the crossings, and this point will be discussed further below.

4.3.3 *Summerland.* The organizational background to the Summerland fire is the most complex and the most diffuse of any in the three cases discussed. The leisure centre was developed by Douglas Corporation, the local authority of the largest town on the island, with financial assistance from the Government of the Isle of Man. The shell of the completed building, which was owned by Douglas Corporation, was leased to a leisure company which had authority to design and build the more decorative part of the interior. There was an important gap in the continuity of the project between the design and construction of the shell under one design team, and the design and furnishing of the building by the lessees, employing a second design team. In addition to its role as developer, the local authority was involved through its planning, engineering and fire safety committees in scrutinizing successive by-law, planning and safety submissions with regard to the building as design and construction progressed.

The design of the shell of the building was placed in the hands of a local architectural practice, which in turn obtained agreement for a larger company on the UK mainland to be retained as associate architects. In the second phase, the design and fitting out of the interior, the British architects were employed as principals by the lessees of the building, the leisure company. Because two of the most important constructional materials being used were novel ones, the manufacturers and distributors of these materials were included within the range of the Commission's enquiries, Additional factors which the Commission considered to be relevant to their investigations were the extent to which informal contacts between those in the island community were developed at the expense of more formal procedures; and the extent to which the pressure to have the second phase of the building completed in time for the tourist season led to the cutting of corners through pressure of work.

A final area of investigation concerned the mode of organization of the staff of the completed leisure centre, in particular the severely limited nature of the arrangements made for the training of staff in fire procedures.

The Commission concluded that the underlying factors were:

> many human errors and failures, and it was the accumulation of these, too much reliance on an 'old boy' network and some very ill-defined and poor communications which led to the disaster [14].

These general factors were operating in a situation in which a small architectural firm was undertaking its first large commission, designing a new kind of building, which posed new kinds of fire risks, and which was built with new kinds of construction materials. In addition, the conditions under which it was anticipated that the building would operate were changed significantly during the design process.

4.4. *The similarities*

The precise patterns of events and even their dominant components are different in each of the three cases outlined above, and this is presumably why parallels have not been drawn in the past. But if we try to look behind these immediate differences, we discern a number of similarities. The common features may be summarized under the following headings. In later chapters, a clustering and synthesis of these features is attempted, but here they are presented in the form in which they initially emerged from the analysis of the three cases:

4.4.1. *Rigidities in perception and beliefs in organizational settings*

The accurate perception of the possibility of disaster was inhibited by cultural and institutional factors. The Aberfan case in particular offers a powerful and tragic instance of the manner in which this failure of perception may not merely be an individual failure, but may be created, structured and reinforced by the set of institutional, cultural or subcultural beliefs and their associated practices.

All organizations develop within them, as part of the equipment which they use in operating organizationally upon the world, elements of continuous culture which relate to the tasks which they face, to the environment in which they find themselves, and to the manner in which those within the organization are to interact both with each other and with any equipment they may use [15]. Part of the effectiveness of organizations lies in the way in which they are able to bring together large numbers of people and imbue them for a sufficient time with a sufficient similarity of approach, outlook and priorities to enable them to achieve collective, sustained responses which would be impossible if a group of unorganized individuals were to face the same problem. However, this very property also brings with it the dangers of a collective blindness to important issues, the danger that some vital factors may be left outside the bounds of organizational perception.

The Aberfan Inquiry makes it quite clear that the pervasive set of beliefs and perceptions within the coal industry was, for very good reasons, oriented almost wholly towards the problems, difficulties and activities of underground mining for coal, and away from tips as being in any sense important for those involved with mining. A whole cluster of factors contributed to and reinforced this set of beliefs and perceptions. Historical and institutional precedent contributed. Tips had been neglected by the 1938 Commission on Safety in Mines, and HM Inspectors of Mines and Quarries had similarly paid no attention to tips. It had never been the practice in the 25 years of the NCB's existence to survey tips, and an international search after the accident led to the discovery that only Dortmund and South Africa had regulations about tips, so that the neglect was not confined to Britain.

The beliefs were expressed and in turn reinforced by the kinds of terms used in discussion within the industry—tips were regarded as discard, and therefore neglected. In the words of the Tribunal:

> Rubbish tips are a necessary and inevitable adjunct to a coalmine, even as a dustbin is to a house, but it is plain that miners devote certainly no more attention to rubbish tips than householders do to dustbins [16].

Whorf [17] has made a similar point about investigations which he carried out as an insurance assessor when he traced the source of some accidents to semantic factors involved in the naming of dangerous objects. At Summerland, a crucial set of stairs was referred to and thought of as 'service stairs' rather than 'emergency stairs' [18] and a description of the project by the architect referred to it as 'not a building, but a weather-proof enveloping structure', which the Commission found to be both confusing and misleading to a client [19].

Individuals within organizations may be thought of as having 'perceptual horizons' with regard to those things which are significant and important to them in the pursuit of their tasks, the positioning of these horizons being influenced and reinforced by institutional beliefs and terms. At Aberfan, for example, the Tribunal commented that:

> We found that many witnesses, not excluding those who were intelligent and anxious to assist us, had been oblivious of what lay before their eyes. It did not enter their consciousness. They were like moles being asked about the habits of birds [20].

When a pervasive and structural set of beliefs bias the knowledge and ignorance of an organization and its members, these beliefs do not merely show up in the attitudes and perceptions of the men and women within the organization, but they also affect the decision-taking procedures within the organization. At Aberfan, little trouble was taken over decisions about where to site a new tip [21] and there was a reliance upon poor, inaccurate information [22]. Additionally they affect the organizational arrangements and provisions. Thus again at Aberfan, no detailed training was given to local men about tips, no civil engineers were appointed at the appropriate levels to deal with tips, and when a civil engineer *was* appointed in 1960, the mechanical engineer looked after tips, in spite of a lack of specialist knowledge for this job. No surveying of tips, was carried out, no record of inspection of tips was felt to be needed, and no notification of tip slips was required. A specific directive to frame a safety policy was not carried out in regard to tips [23], and warning signs of instability were not reported to the appropriate specialist when they did begin to occur [24].

Thus, there is a possibility of a vicious, self-reinforcing circle growing up where it is generally believed that an area is not important or problematic [25]. The staff employed, therefore, are not specialists in this area, and are not employed to look at it, so that the presence of new staff serves to reinforce the beliefs which placed the new staff there.

4.4.2. *The decoy problem*

There is another important recurrent feature in the reports analysed: in a number of instances where some hazard or problem was perceived, action taken to deal with that problem distracted attention from the problems which eventually caused trouble.

Or, to rephrase this in the terms used earlier in this chapter, a contributory factor to both the Hixon and Aberfan disasters could be considered to be the attention paid to some well defined problem or source of danger in the situation, which was dealt with, but which distracted attention from a still dangerous but ill structured problem in the background. To cite some examples: the Tymawr tip slip in 1965 led to a general acceptance of the slip potential of tailings (very fine waste), which misled the corrective action taken at Aberfan. When the tipping of tailings at Aberfan stopped, fears about a slip were allayed [26]. Similarly, one of the local council's complaints about danger was linked with a proposed aerial ropeway scheme for tipping; when the scheme was withdrawn, the complaint was withdrawn [27]. Summerland was not seen as a conventional large building or as a conventional theatre, and the appropriate regulations were not, therefore, applied.

At Hixon, a director of the haulage firm stated that he realized the danger of the new crossing with regard to arcing of lorries on to the overhead wires, or with regard to lorries stalling on the crossing, but not with regard to the danger of crossing slowly [28]. Similarly, an English Electric manager checked the crossing with regard to the danger from arcing, and the police escort warned the driver of the long vehicle carrying the transformer of the dangers of a hump across the track, and of the overhead clearance [29], and a meeting was held to discuss the dangers from arcing when transformers were carried across the crossing [30]. Again, in a slightly different context, with a different well structured problem, British Rail was consulted in 1967 about the dangers of abnormal loads, but their engineers interpreted this entirely in terms of possible damage to bridges over which the load might travel. In this case, the Bridge Engineering Division represented an institutionalized response to the problem of bridge damage and safety, and this institutionalization of the one problem further inhibited the perception of the 'slow-moving load across automatic half-barrier crossing' problem [31]. Publicity surrounding the introduction of the new crossing was directed at children, with much less adequate coverage of adults, whereas, in fact, only adults were involved in the events leading to the accident.

An initial response to the decoy phenomenon is to consider whether there may not be means by which organizations can scan the problems which they do in fact deal with, and then try to look behind them to see if they are obscuring potentially dangerous ill structured problems. This may possibly be a successful strategy, but there is a problem of infinite regression, since the

THREE DISASTERS ANALYSED 61

moment a new problem becomes clearly defined, the possibility arises that it may be obscuring some further problem. A way of seeing is always also a way of not seeing. However, this does not automatically rule out the possibility of taking some precautions in this direction, and the topic is returned to below.

4.4.3. *Organizational exclusivity: disregard of non-members*

In two of the cases under discussion, individuals outside the principal organizations concerned had foreseen the danger which led to the disaster, and had complained, only to meet with a high-handed or dismissive response from the organizations concerned, the National Coal Board [32] and British Rail [33] respectively. The 'fobbing off' of Aberfan Council with ambiguous and misleading statements such as 'We are constantly checking these tips', and the use of crossing site meetings as public relations exercises by British Rail both indicate an attitude that those within the organization knew better than outsiders about the hazards of the situation with which they were dealing.

There is a dilemma here: for, on the one hand, a task may be difficult to accomplish if too much time is spent on listening to complaints from outsiders. But, on the other hand, the tendency discussed below to deny remote dangers makes it easy for the administrator to label as 'cranks' those who fail to agree with his organization's policy.

4.4.4. *Information difficulties*

Information difficulties are likely to be associated with ill structured problems, since it is not easy for any of the individuals or parties involved to fully grasp and handle these vague and complex problems. In situations of 'disjunct information', the simple remedy of 'better communication' will not work unless resources are increased so that the problem is no longer ill structured, or unless the problem defined is reduced to a size which can be adequately handled by the existing information net [34].

It seems likely that communication and information-handling difficulties are widespread in all organizations, so that it would be wrong to suggest that all such cases lead on to disaster. However, although the kinds of disaster being examined here were chosen particularly because they were likely to display information difficulties, the variety of such problems is quite surprising. The manner in which such problems might be related to the subsequent emergence of disaster is considered in discussions in later chapters.

Communication difficulties were referred to in each of the three reports examined, with the following summary of the situation within the

Ministry of Transport being made in the official Hixon report [35]:

> The number of officers within the Ministry who have been concerned with the matter over the years have been far more numerous than those who are now available as witnesses. Moreover, the Ministry consists of a number of large departments, each of whom may gain a piece of knowledge which, added to what another department knows, might produce realisation of a particular fact, but it is sometimes inevitable, to use the words of one witness, that 'with the best of intentions on the part of the individuals concerned, something is likely to fall between the interstices of the administrative net'. Unless the amount of paper in the Government service is infinitely increased, it is not practicable to prevent such a mishap absolutely.

To try to pull out in a more analytic manner some of the issues raised in this quotation, it may be noted that it is characteristic of the class of problems under discussion that they are being handled by many actors, each with his or her own 'theories' about the nature of the situation in which they find themselves, and often with a considerable degree of discretion in their actions. Since the problem is ill structured, responsibility for handling it is often distributed in a vague or unclear manner. At Summerland, responsibilities were blurred by an important gap in the continuity of the project, between the building of the shell of the centre by one team, and the fitting and equipping of the building by another (though related) team [36]. Other ambiguities arose around the control of the adjacent building [37], and around the functions of the planning Committee which handled planning applications [38].

For similar reasons of lack of knowledge, lack of time, lack of realization of the importance of the issue and conflicting demands, the *orders* issued to those who are concerned are often vague and imprecise. Thus, in connection with Aberfan, a National Coal Board Divisional Chief Engineer assumed 'that there would be detailed collaboration between colleagues in checking tips, and the absence of the involvement of mechanical and civil engineers in this was literally disastrous'. Several such patterns of ambiguous instructions occurred also at Summerland [40], where the following situation obtained before the fire:

> Mr De Lorka thought it was for Mr Harding to organise an evacuation procedure, but he never discussed it with him ... Mr Harding thought it was for the heads of departments to organise their own evacuation procedure but he gave them no instructions about it. Mr Paxton, the Deputy Managing Director of Trust Houses Forte Leisure Ltd, thought it was for Mr Harding to organise an evacuation procedure and for Mr De Lorka to make sure that he did it. Mr Dixon, the supervising Fire and Safety Officer of Trust Houses Forte Leisure Ltd ... thought it was Mr De Lorka's duty ... and no part of Mr Harding's duty. Mr De Lorka, in evidence, accepted that if a fire oc-

curred, he relied on members of staff using their own initiative as to what to do to get people out safely ... [41].

Again, British Railways publicity about their new crossings gave ambiguous instructions to drivers of heavy vehicles [42]. A particular form of this failure to give precise orders which is noted below is the attempt to avoid the need to inform selectively the parties concerned with a complex problem—by sending *all* the information to all of them, as with the technical memorandum on automatic crossings set out in 'otiose paragraphs' by British Railways and circulated without differentiation to BR technical staff, the Ministry of Transport, the Home Office, the Police, the Magistrates' Association, the Royal Society for the Prevention of Accidents and the Road Hauliers' Association.

The position is complicated because individuals have their own theories and views about how their information is likely to affect the total situation, and they generate their own rules for action from these views. Thus Colonel Reed, of the Ministry of Transport Railway Inspectorate, drew a clear line around the limits of his concerns and responsibilities, denying that he should have been concerned with detailed traffic conditions, a denial which, with the benefit of hindsight, the Inquiry did not accept. He stated:

> If we are going to think along these lines, then any level crossing in relation to any built-up area would have to be specially considered in regard to the industrial use generally throughout the area ... If we are going to say that, at Hixon, we ought to have thought about the airfield, then I would say that at Hensall we ought to have thought of the long-distance traffic which we know used it regularly ... and at Nantwich, Cheshire on a busy road, we ought to have thought of all the traffic that might use it, and there is a limit to this [43].

A particular form of conflicting view of the nature of an ill structured problem which occurred in both the Hixon and Aberfan cases was the adoption by some top management group of an idealistic and unrealistic view of the problem area. It is relatively easy for a managing director or a chief executive to describe his view of how his men or his departments are coping with all eventualities and considering things from all angles, because he is remote from the area which he believes is covered by safety regulations [44], and unless his comments are followed up they are unlikely to be tested by reality except in the case of disaster. The top management of the NCB had an idealistic view of what should have happened with regard to tip safety and tip siting, but they took no action to check on it, or to check that men with the right training were available [45]. Similarly, the directors of the haulage firm at Hixon had a policy of leaving many things to their drivers, who had a large degree of autonomy, so that, as was commented 'In this state of ignorance and failure to draw the glaring inferences from the events which had taken place in their own business' they failed to instruct their drivers or to warn them about automatic crossings [46].

A further difficulty, which was discussed at length above, is that people may be inhibited from communicating about the problem which eventually causes a failure, because their attention is fully occupied in dealing with more clearly defined problems. These problems, which may be important ones in themselves, act as 'decoys' to draw their attention away.

With all these contributing preconditions, then, what forms of malcommunication actually occurred?

Occasionally, *wrong or misleading information* was sent [47] or information was *sent to the wrong people*. Thus, a report on tip safety did not reach NCB Headquarters, or Aberfan: and British Railways' instructions about crossings were not sent to the BR division responsible for abnormal loads.

Information may be *distorted in transmission* and a more subtle failure occurs when information which the recipient has expected to be processed, or helpfully distorted, is passed on untouched. Thus, the Home Office, out of deference to a Civil Service practice of not interfering with information from other Departments, failed to do the staff work on which the police had come to rely, on the Ministry of Transport automatic-crossing memorandum [48].

Information transfer may also be inhibited because of poor communication between two particular individuals, arising from personality or other differences. A closely associated phenomenon would seem to be the emotional response which may accompany or be provoked by communications between individuals or groups when these do take place. As will be noted below, both British Railways and the National Coal Board on occasions communicated with people outside their respective organizations, about the problems which eventually led to accidents, in a manner which provoked responses of anger and annoyance.

Another form of information distortion or diversion is that which arises when too much reliance is placed upon informal networks (established for other purposes) to communicate about complex problems, as in the case of the over-reliance on the 'old boy' network at Summerland.

A further source of difficulties with information-transmission, related to ill structured problems, lies in the *existence of ambiguity*. Often, because of the vagueness and complexity of the problem, ambiguity is difficult to avoid, whether it is ambiguity about the evidence available to be transmitted, or ambiguity about the roles and motives of other actors involved in the situation. Thus, at the Aberfan tribunal inquiry, there were a number of disagreements about the nature of earlier slips. At Summerland, considerable ambiguity existed about the characteristics of the acrylic panels when exposed to fire [49]; and when a routing section within the Ministry of Transport was informed of the existence of a new automatic crossing at Hixon, they reported that they spent some time considering the question of 'exactly what category of information this particular feature came into' [50].

When information has been transmitted and received, the information-handling difficulties are not over, for the recipient may well fail to deal with the problem to hand, even with all the information available, for a number of reasons:

(a) The relevant information may be buried in a mass of irrelevant information. When selective communication is evaded by sending all the information available to all parties, the recipient has to adopt strategies to avoid being overwhelmed by it. The information may be overlooked [51], or the recipient may be busy and resist absorbing other material, particularly when he judges the incoming item to be irrelevant to his current concerns: the police constables at Hixon had disregarded the automatic-crossing memorandum, which was available in every police station, as 'mere flotsam in the station' [52]. The overlooking of information becomes more serious when, as at Summerland, the conditions which create a pressure of work are the very conditions which heighten the danger [53].

(b) The recipient may fail to attend to the information because it is only presented to him at the moment of crisis [54].

(c) He may adopt a 'passive' mode of administrative response to the issue. When an individual receives communications which he regards as not central to his main concerns, his response may be to avoid taking any action with regard to such peripheral matter unless comment is actively invited. Thus, communications which are sent under the heading 'for information' are sometimes not treated as information at all. We could characterize this more generally as an instance of a 'passive mode of administration' which follows a philosophy of 'don't look for trouble'.

(d) He may fail to put the information together creatively. The suggestion that a passive type of administrative response occurs draws our attention to the possibility that its counterpart, a more active response to administrative problems, may also be encountered. The absence of such an active response was an issue which occupied the Hixon Inquiry, and much of the evidence of the Inquiry related to the conditions under which a succession of individuals, who had all of the information potentially available to them, failed creatively to assemble the possibilities of failure in a clear enough manner to lead them to act to prevent it.

Thus, the hauliers did not see the danger even after one of their lorries had grounded on a crossing at Leominister: they did not 'put two and two together' because the vehicle at Leominster was a small one, not a long, abnormal load [55].

The British Railways instructions to signalmen about heavy loads were known to the Railway Inspectorate, but this 'did not bring the problem to their minds' [56]. The risk was not hidden from the Bridges Engineering Design Standards Division but no-one worked out the time necessary for a 150-foot vehicle to cross [57], and the problem was not seen any more clearly in the Ministry of Transport than elsewhere. A number of the witnesses had evidently been troubled by this failure, a concern which shows in their evidence. Thus, Mr Scott-Malden, Under-Secretary in charge of the Railways Group, referred to 'an omission resulting from lack of imagination', when pointing out that the Railway Inspectorate and the Bridges Engineering Design Standards Division possessed, before January 1968, all the knowledge necessary to anticipate or foresee such an accident as happened, but though their officers were competent and intelligent men, no-one foresaw precisely the nature of the problem. He said [58]:

> I think the two pieces of knowledge ... had to come together in one person's mind, and he would have to see the connection between them. That is what could have happened really anywhere in the Ministry or indeed in quite a lot of other places. But that linking, and that, as it were, flash of imagination did not happen.

Failure to take positive or creative administrative action also arose at Summerland, where the Commission noted that no-one, clients, authorities or architects, ever stood back and looked at the project as a whole, although each could have done so within the terms of their responsibilities [59].

Such conditions raise very clearly the issue of creative problem-solving and anticipation in administration, and suggest that one avenue which might be explored in the attempts to avoid future disasters is the stimulation of these kinds of activities among managers and administrators, perhaps using the existing literature on problem-solving as a mental activity [60].

This issue becomes more acute as larger organizations and more complex problems increase the possibility that the decisions leading to major accidents are taken by those remote from the scene at which the accident is likely to take place, so that their normal response to imminent danger may be dulled [61].

A final form of information difficulty which is encountered in the cases studied is the difficulty in transmitting adequate amounts of information (about appropriate action and precautions in hazardous situations) to members of the general public or to what we call, below, those in the category of 'strangers'.

4.4.5. *The involvement of 'strangers' especially on complex 'sites'*

When potentially hazardous procedures are being carried out by individuals there is a need to ensure that these people know the best available means of coping with the potential hazards. Where a small, clearly defined group is concerned, and especially when they are employees of one organization, the problem of giving them adequate information is relatively simple. Thus, operators of dangerous machines in a factory can be carefully selected, trained and instructed. When those who may activate the hazard include a wider class, the class of all members of some organization for example, or all members of a particular department or rank in an organization, the same problems of selection and training occur. In addition, though, since individual members of this large class of people are not likely to be very frequently in the hazardous situation, the problem arises of giving them information which will remain with them in spite of conflicting demands upon their attention, and will be available for use when they need it.

This issue became particularly important at Summerland where two classes of people became involved in the disaster, neither group being adequately informed. The public did not receive adequate instructions about what to do because the staff failed to use the public address system, or to instruct them adequately in other ways. And the staff themselves had not been instructed about how to behave in the event of a fire occurring. Rules and instructions did exist, but this is clearly inadequate if the staff do not know them and are not encouraged to follow them. As the Commission commented with regard to fire-exit notices which made little impression upon staff or public: 'A proper evacuation system is not established or maintained merely by putting up notices' [62].

There is a further problem of limiting access, thereby ensuring that only members of the 'informed' group can place themselves in the hazardous situation, which means that entrances must be controlled to prevent others placing themselves at risk. But hazardous situations also exist when those not directly under the control of any of the organizations concerned can put themselves in positions where they can activate the hazards by behaving improperly. This group will often, though not always, be members of the general public, and it seems useful to refer to them as *strangers*.

The basic problem which exists with regard to strangers is that they are difficult to brief, because as a group they are difficult to define. Thus, information about the desired safe procedures must be disseminated to a wide amorphous group of potential users, many of whom will never actually need it, or will not know that they need it at the time. In practice, this problem is tackled by mounting publicity campaigns, and by placing warning notices and instructions, both of which were deficient, for example, at Hixon [63].

The 'strangers' who did encounter information about the Hixon crossing were also misinformed to an extent—the material available in 1967 was out-of-date [64], no mention was made of slow-moving vehicles [65], the message was not urgent, it did not stress the danger of the crossings (unlike the Dutch publicity film *One Minute to Eternity*) and it was misleading with regard to the approach of a second train. Too much publicity was felt likely to make the public too anxious about the crossings. In addition, the formal 'site visits' arranged before crossings were opened were seen, in part, as public relations exercises by those conducting them, designed to allay public misgivings rather than to publicize accurate information; at the Hixon site meeting, in consequence, no mention was made of the risks attending the use of the crossing, or of the availability of the telephone.

When the Ministry of Transport decided that automatic half-barrier crossings were 'self-evident hazards' like traffic lights, they were making assumptions about how they would be perceived by the amorphous body of strangers known as 'road users', and such assumptions are evidently dangerous ones to make. At Summerland, the stereotyped view of the public and its likely behaviour in the case of a fire ignored those parents who were separated from their children in the children's cinema on another floor. These parents fought to reach them against the flow of the crowd, increasing congestion on a crowded and dangerous staircase [66]. The Summerland fire highlights the special problems posed for those responsible for hazardous situations where safe operation relies to some extent upon the safe behaviour of strangers. These problems are intensified by the fact that the strangers are always, in these cases, on the site when correct behaviour is demanded of them; and since they are in the concrete situation, they have a variety of opportunities to manipulate the situation in manners not foreseen by those designing the abstract safety system: they may find other uses for objects on the site—throwing life-belts into the canal, cutting up signal wires to sell for scrap, spraying notices with aerosol paint, and so on.

Any concrete or material system possesses many properties, all of which are potentially evident when the system is directly encountered [67]. Even though any one object may be present in a system because of only one of its properties, it unavoidably brings with it all of its other properties. Thus, as Pask has commented, a thermostat may be representable as an electrical systems component. However, the circuit diagram does not show the mechanical properties of the thermostat which come along with the electrical ones. 'On site' it is possible for me to stop a thermostat from operating in a manner which is totally unpredictable from an inspection of the circuit diagram—by kicking it [68]! Human ingenuity is endlessly resourceful in finding ways of manipulating the objects in a concrete situation in a manner unforeseen by the designers of one abstract aspect of that situation.

The materials used in constructing sites have multiple properties, some of which may be ignored or overlooked, as when a plasterboard wall at Summerland was replaced by a sound-absorbent fibreboard wall, which was also combustible [69]. And even when safety features are designed into a system or situation, the safety features themselves have a multiplicity of properties which may complicate their operation. People may not realize that a particular device *is* intended to contribute to safety [70]. Safety devices may function in many ways: emergency exits may also offer opportunities for unauthorized entry, so that it is not only at Summerland that fire exits have been found to be padlocked when a fire has occurred. Moreover, safety features may also have affective properties which influence the manner in which they are used. Wolfenstein has pointed out that air-raid shelters may induce fears of darkness, or of claustrophobia, and for those reasons may be avoided. Alternatively, they may be regarded positively because of their sociable features [71].

Any problem which is concerned either to create a site (planning, town planning, hospital planning, level-crossing planning), or which uses sites, becomes involved at that site in a multiplicity of systems, some designed, some unpredictable. For this reason, except on the smallest and most accessible sites, any problem involving a site is potentially an ill structured one: the design of subways as means of separating pedestrians and traffic overlooks the opportunity created 'on site' for these subways to be used by teenage gangs or by exhibitionists. The design of lifts as a means of taking tenants up to the top of high-rise flats overlooks the creation of a social no-man's-land which becomes perilous to enter, and of which no-one will take care.

Only when the site is small enough for the implications of all of the concrete variables to be considered is the production of such undesired consequences likely to be avoided [72].

4.4.6. *Failure to comply with regulations already in existence*

This heading evidently covers an extensive class of phenomena, which was particularly important in the Summerland case, and to a lesser extent at Hixon. For reasons set out in a previous section, few relevant regulations appear to have existed in the Aberfan case.

(i) *Failures of submission*. Some of the breaches of regulations occurred at Summerland because *inadequate plans were submitted*, sometimes with a feeling of 'we might get away with it', or 'take the regulations with a pinch of salt' [73]. The plans may be inadequate in that features are not shown (e.g. the rustic stair at Summerland) [74], information is not given (e.g. the application for the theatre licence at Summerland) [75], or modifications are not shown [76]. Alternatively, those concerned may be *unaware of the regulations* [77] or may not realize that they apply to the case in hand [78].

Those proposing to take actions not in accordance with the regulations may conceal this fact, of course, or they may have the intention of applying for a waiver, though may not follow this intention through [79].

(ii) *Failures of control and implementation*. There may be difficulties in applying regulations because of changing technical, social or cultural conditions which make existing regulations inapplicable. Thus, at Summerland, a number of comments were made about the difficulties of applying conventional Theatre Regulations to the new buildings [80], leading in turn to unsuccessful attempts to apply to the new situation what was felt to be the spirit of the Regulations, or, alternatively, to formalistic or ritualistic defences of substantive breaches of regulations [81]. When waivers of regulations are being allowed, it may not be possible to control finely enough the conditions or the partial nature of the waiver [82], so that sometimes applicants may be given more than is intended, or may get more than they requested [83].

Implicit in some of the examples discussed above, is the notion that failures occurred because of *failures of controls on controls*. Often these are failures of inspection [84], the inspector being the 'controller of controls', but they may also occur when rules applying to the amendment or waiving of regulations are not complied with.

THREE DISASTERS ANALYSED   71

The Summerland Commission felt that the waiver of a safety provision is always a responsible decision and that a design on which a waiver is permitted should incorporate some compensatory measure to restore the standard of safety to that which would have existed [85]. They also pointed out that there was no clear, prescribed procedure for the handling of waivers [86] and that there was a long-established failure to observe the provisions of the Act applying to waivers [87].

For similar reasons, the Summerland case would suggest that attention should be paid to the position, powers and responsibility of those responsible for giving waivers, for at Summerland these powers were delegated to someone with insufficient authority to exercise them properly [88].

4.4.7. *Minimizing emergent danger*

Another problem which recurs at many points in the three reports is that of failing to see or fully to appreciate the magnitude of the danger of some potentially hazardous situation. We may distinguish a number of states in the progression from this failure of perception to a full realization of the danger, displayed by individuals in the three reports studied.

(i) *Underestimating possible hazards.* In the earliest state, *some realization* of the possible complexity and danger of the situation is achieved, *but this is seriously underestimated* [89]. At Aberfan, the very visible evidence of previous large slips was ignored [90], and little trouble was taken over the siting of tips [91]—'the blind leading the blind in a system inherited from the blind'. Before the Hixon incident, the Leominster incident mentioned above 'ought to have made the directors of the haulage firm aware of the problems of automatic crossings' but it did not [92] and British Railways saw slow vehicles as a potential hazard, but not as a possible cause of a serious hazard. Again, before the Summerland fire, although some risks from fire were acknowledged, waivers were given on the use of inflammable material in the building and exemption was allowed from some aspects of the Theatre Regulations.

Even where safety precautions are taken, they may be used more for their talismanic properties in warding off danger, or as a propitiation of the 'powers-that-be', than as serious attempts to prevent an incident occurring. Thus, in discussions at Summerland about why the service stairs were not adequately protected against fire, the architect (in evidence) made the 'rather startling observation that this staircase was "a notional fire escape at the time ... an earnest of intention"' [93].

Given the complexity and vagueness of ill defined problems, it is often difficult for one party with one view of a hazard to convince another party of the validity of that view. Characteristically, we therefore find a tendency to undervalue or ignore the conflicting diagnosis of a dangerous situation offered by other groups. The National Coal Board's response to complaints and expressed fears from Aberfan residents was to ignore them or to assume that there was little substance to them [94]. Similarly, British Railways failed to realize the sound basis of the fears expressed by the directors of the haulage company after the Leominster incident, and brushed them off, as they did the fears of the public about the new forms of crossing [95].

A whole range of possible strategies may be adopted by individuals dealing with an ill structured problem, when confronted with other people's conflicting beliefs. They may devalue the problem, so that any conflict which does emerge is seen as unimportant: they may fail to follow through clearly the implications of conflicting views; they may develop rationalizations, which explain away the other parties' views [96]; or they may tacitly ignore or avoid contact with the kinds of information which might force them to reappraise the situation, as the National Coal Board personnel collectively ignored the existing literature on tip safety measures [97]; or they may handle the conflict, as we noted, by making statements about the problem area which seem to indicate agreement, but which in fact are ambiguous. A number of examples of this appear in the Aberfan evidence: the statement that the position of the tips was under constant check [98]; the use of an ambiguous plan for a check on tip dispositions [99]; and the situation in which the council complaints about tip safety became attached to a somewhat enigmatic plan for an aerial ropeway, so that when the plan for the ropeway was dropped, the council relaxed its vigilance.

Hazards may also be underestimated in ill structured situations if one party forces acceptance of its definition of the level of hazard, by the use of power which may derive from control over resources, from status, from an 'expert stance', or from the ability to impose secrecy in crucial areas. A relevant example of this is provided by the Aberfan Tribunal's comment on the National Coal Board's 'subterfuge and arrogance' in dealing with the residents' complaints, and the stubborn efforts of the Board to resist attempts to lay blame at their door even after the disaster, the Tribunal noting a lack of frankness in their admissions [100].

Similarly, when Wynn's lorry grounded on the track at the Leominster crossing in front of an approaching train, a railway official's comment to the driver was 'You can't park there!' The British Railway's letter to the haulage firm about this incident, a letter 'remarkable for its arrogance and lack of insight ... at a high executive level in British Rail' [101] stated that vehicles must not become immobile on crossings [102].

THREE DISASTERS ANALYSED

(ii) *Minimizing emergent danger.* In the next stage to be considered, the *possibility of real danger emerging* begins to occur to some of the individuals or groups concerned, *but many of those concerned seem to undervalue it* [103]. Why should this be so? Are there pressures which encourage those in specific organizational settings to undervalue emergent danger [104]? Is it the fear of being called alarmist? Or is a stronger factor the asymmetry whereby anyone taking action must weigh up the remote possibility of a catastrophe against the immediate and certain outcomes of taking action to avert that catastrophe? Wolfenstein [105], noting a pattern of denial of danger, relates it, in an extended discussion, to psychological factors leading to the repression of painful ideas, or to the isolation of emotion from an idea which is accepted cognitively.

(iii) *Conflicting views about danger.* When conflicting views about danger are held, the individuals concerned may choose to take time to review the information used and this is often done by visiting the site in question to check up on the situation at first hand. But there are limits to the extent to which site visits provide an efficient means of agreeing about the hazards of a situation; for though, in some cases, differing views may be reconciled at a site visit, in other cases they may still remain separate, so that different participants see different things.

Visits were made to the Aberfan tip and to the Hixon crossing. As already noted, a 'site' represents the concrete aggregation of whatever abstract systems have been imposed upon it, together with an amount of ancillary and fortuitous materials (and people). At boundaries or interfaces between systems, or in complex systems, several types or levels of abstraction may be superimposed. Thus, the road/rail interface at level crossings of the Hixon type contained features not planned for, features which challenged the organizational perceptions of British Railways and the Ministry of Transport, and which raised so many issues that, as we have seen, a full exploration of the complexities of the events possible at a crossing site was ruled out by Colonel Reed, even after the Hixon accident had occurred.

(iv) *Changed awareness of danger.* When the full scale of the possible danger is finally realized, the apparently straightforward response of admitting this and taking the appropriate precautions does not always occur. A defensive attitude may be adopted by one group, or others may begin to take steps to absolve themselves from responsibility [106]. Attempts may be made to control the situation, Canute-like, by fiat—'Vehicles must not become immobilized on these crossings' [107].

(v) *Failure to call for help.* Finally, when individuals are exposed to danger, it seems to be not uncommon that they do not call for help immediately. At Summerland, the staff failed to call the fire brigade promptly, and the elaborate fire-alarm system was not used at all [108], one of the first warnings of the fire being given by a ship at sea which spotted the blaze on shore.

This pattern is not limited to Summerland, for Barlay [109] cites a study of 1200 fires, one sixth of which had become large because of a failure to summon the fire brigade. Again, questions are raised about whether such behaviour occurs because of a fear of sounding an unnecessary alarm, or because of a persistence of the syndrome of denial of danger, a persistence which Wolfenstein [110] suggests becomes more pathological the nearer the danger becomes.

4.4.8. *Nature of recommendations after the disaster: the definition of well structured problems*

An important function of most committees or tribunals of inquiry is the making of recommendations or guidelines for formulating policy to prevent recurrences of the kind of disaster being investigated. Thus, the Aberfan Tribunal recommended guidelines of good tip practice, the formation of a Tip Safety Committee, the establishment of a Code of Practice relating to tips, the employment of Inspectors, and the training of men and managers in aspects of tip stability. The Hixon Inquiry recommended that more publicity should be employed, crossing widths should be limited, the warning delay should be increased, more warning signs should be provided, more severe penalties should be imposed on those behaving improperly at crossings, signals and crossings should be redesigned, and mandatory instructions about the use of the telephone should be issued to drivers of abnormal loads, and those concerned with them. The Summerland Inquiry recommended, among other things, the revision of the Theatre Regulations and the provision of more training in fire regulations for architects.

All of these recommendations, diverse though they may be, have in common the following: that they are concerned to deal with the problem which caused the disaster as it is now revealed, and not to deal with the problem as it presented itself to those involved in it before the disaster. As the Summerland report phrased it [111]:

> It would be unjust not to acknowledge that not every failure which is obvious now would be obvious before the disaster put the structure and the people to the test.

In terms used earlier in this chapter, the recommendations are designed in general to deal with the well structured problem defined and revealed by the disaster in question, rather than with the ill structured problem existing before the disaster. In part, the intention of this book is to consider whether there may be some general principles which could be formulated to deal with at least some 'ill structured' problems *before* they are sharply defined by disaster, to seek an understanding of the origins of disaster.

4.5. *Discussion of the analysis*

The analysis above, based as it is upon an examination of only three cases, is exploratory in nature; but, by bringing together all of the factors reviewed in the cases analysed, it can be tentatively suggested that the kinds of conditions which might very well provoke a disaster would be some combination of the factors set out below. This summary may help to clarify the choices which face administrators and managers concerned to avoid disasters, although it is offered with an all-too-painful awareness that many of the possible courses of action discussed could more readily be represented as dilemmas for administrators, rather than as clear-cut guides for action.

4.5.1. *Factors which may combine to produce disaster* [112]

(*a*) an inter-organizational grouping of one or two large organizations and some smaller ones involved in
(*b*) a complex, ill defined and prolonged task which gives rise to
(*c*) a variety of information difficulties.

During the course of this project:

(*d*) its goals are likely to shift,

and because of the prolonged nature of the task

(*e*) the administrative machinery concerned with the task is likely to undergo changes, and
(*f*) some of the parties will change their roles in relation to the task.

In any case, because of the complex and ill defined nature of the task:

(*g*) there will be ambiguities associated with the handling of it;
(*h*) regulations relating to the task may be somewhat out of date, or may not be stringently enforced.
(*i*) The individuals working in the area are preoccupied, by virtue of their organizational positions and professional or occupational background, with some major issues relating to the task in hand, and they are reinforced in their preoccupations by organizational tradition and precedent.

A characteristic task for such a grouping is:

(j) the design of a system which includes large or complex sites
(k) to which the employees of a number of organizations have access, and
(l) to which the public is also admitted.

The members of the organizations concerned operate, in their official capacities at least, with

(m) stereotyped views of the public and its likely behaviour with regard to their project.
(n) Complaints from the public are usually treated in a fairly cursory manner, since they are felt to come from non-experts who do not fully understand the issues involved, and who do not have access to all the relevant information. Sometimes with justification, it is pointed out that such complaints are made by unduly nervous cranks.
(o) Where signs of possible hazards emerge, some of them will be recognized and planned for.

Others will be neglected:

(p) because they are not recognized by those operating with approved organizational stereotypes;
(q) because of pressure of work;
(r) because recognizing them and taking action would call for the investment of time, money and energy in courses of action which would be difficult to justify within the organization; and
(s) because most of the individuals concerned feel that quite probably it won't happen anyway.

4.5.2. *Possible lessons at an early stage of the analysis*

In examining this list of factors, we may ask whether there is any way in which it can be used to point out possible ways of preventing in the future the failure to cope with other ill structured problems which are still in existence. The specific kinds of well structured problems revealed by the disasters have been dealt with by the Tribunals and Public Inquiries. Ultimately, of course, no-one can offer a certain way of avoiding all pitfalls, but some possible precautions may be suggested. It may be that the precautions will be too costly for the potential benefits they offer, or that they will be found to be too restrictive, since human beings have to accept and cope with some degree of risk in their lives. But as society deals with and becomes dependent upon larger organizational systems, the possibility of someone who is devising a system taking a risk which leads to disasters for many others who were not parties to the risk-taking decision becomes more and more likely [113], and it becomes important to explore 'the conditions which foster the failure of foresight.'

We began from the assumption that *information difficulties* associated with ill structured problems are not likely to be easily solved, and may even be insoluble. In considering possible actions to deal with such difficulties it is an advance to recognize that perfect, complete communication is not likely to be attained, for this makes it possible to consider which patterns of selective communication it is desirable to encourage. Existing patterns of communication can be examined, and persistent peculiarities of communication scrutinized to see if their persistence can be linked with some aspect of the ill structured problem which is not immediately apparent [114]. Information difficulties which seem to arise from a tangling of communication with emotional issues also deserve attention, particularly when hazardous issues are concerned.

Other informational problems may be listed as common managerial problems, for which little can be suggested as solutions beyond the normal managerial remedies.

These are the problems of obtaining adequate intelligence, of avoiding transmission of the wrong information, of avoiding its dispatch to the wrong people, avoiding distortion in transmission, avoiding the failure to operate on messages when this is expected, not relying too much on informal networks created for other purposes and avoiding ambiguous communications.

Even when an individual in an organization has received all of the necessary information about a problem, we noted above that he may fail to deal with this information for a variety of reasons. Since the recipient may be swamped by too much information, those attempting to communicate should beware of avoiding their responsibility for selective communication by adopting the expedient of providing all of the information, instead of looking at the problem from the recipient's point of view, and attempting to tell him what it is thought he needs to know.

Managers and officials have little alternative than to behave in the 'passive administrative mode' with regard to many of the communications they receive; but if documents are *sent* 'for information', what attitude and what action is expected on the part of the recipient? And if documents marked 'for information' are *received*, what criteria can managers and officials adopt in dealing with them?

The absence of a creative response to the assembly of information already available is an issue which was of great importance with regard to the Hixon accident, and it raises the problem of how individuals can be encouraged to make such a response in an administrative setting. Again there are, of course, limits to be noted, for no official can behave creatively and actively with regard to every topic with which he deals, but some possible courses of remedial action do suggest themselves.

The first step seems to be achieving an awareness that such a positive response is part of a manager's job. Secondly, organizational opportunities may be provided for the positive examination of a manager's job-situation and problems, encouraging his attempts to make a creative response. Thirdly, this problem raises the important issue of ways of dealing with organizational bias and decoy phenomena. No permanent solution is possible here because of the inevitability of infinite regress noted above, but it seems worthwhile, none-the-less, to ask members of an organization, individually and collectively, to set out those well structured problems which their organization is designed to cope with, and then to ask them to try to look behind and around these problems for some closely related problems which could cause trouble but which they have so far ignored. The role played, or assumed to be played by other organizations, is particularly worth looking at here. A similar exercise could be used in trying to locate and deal with problems arising as a result of organizational or industrial bias. When problems have emerged, they can be evaluated, and suitable attempts made to deal with them.

One major obstacle in such an exercise lies in the structured perception and 'blindness' which members of any organization acquire as a result of their membership of that organization, and for this reason, information and perceptions from outside the organization are particularly valuable sources of insights and new perceptions. A possible strategy which could be considered here is paying careful attention to the complaints received from members of the public outside the organization. Of a sample of complaints, it could be asked: 'Suppose this person is right. Suppose that though this complaint may be emotive, ill informed or biased, it contains a grain of truth. Can we deal adequately with the point raised?' The sympathetic treatment of complaints may enable them to be used as sources of alternative perceptions, rather than as a source of topics for public relations exercises.

A similar exercise could be attempted by carrying out a 'hazard audit' in which emergent dangers are re-evaluated. In such an audit, we could ask: what dangers and hazards are we aware of? What grounds do we have for assuming that every one of them might not assume catastrophic proportions? Are our assessments 'realistic', or are they 'normally optimistic' in a way that places others remote from us in danger? If the re-evaluated problem still seems remote, our original stance is confirmed, but if not, more adequate provision may be needed to deal with the issue. Top management in particular might carry out this kind of exercise with regard to the hazards with which they think their subordinates are dealing.

With regard to *safety regulations*, the cases studied suggest a need to make sure that regulations are known by those concerned with them, and that they take them seriously. It seems to be particularly important that those in 'control roles' or 'inspection roles' carry out their duties adequately; where waivers are given, the ground conceded needs to be carefully scrutinized, and the principle of 'compensation for reduced safety' applied. Procedures for making waivers should be clear, and those administering them should have an authority appropriate to their responsibility. In all considerations of safety regulations, sight of the ambivalent psychological attitudes which many individuals have towards such regulations should not be lost.

Finally, two substantive areas of organizational operation have been highlighted in the earlier discussion: *strangers* and *sites*.

Assuming that all personnel within an organization have been adequately trained with regard to the hazards they might encounter, particular attention needs to be paid to any operation which involves 'strangers'. These may be the general public or less obvious strangers: people who work for the official's organization but not for his department, who may wander in looking for someone—delivery men, inspectors, policemen, gas or electricity service men, cleaners, etc. Does safety depend on them behaving 'sensibly' with regard to any of the operations in the organization? How do they know how they should behave? How effective are the methods of telling them? How is this effectiveness assessed? How can the organization deal with failures, ignorance, forgetfulness, stupidity, mental deficiency on their part? If it is intended to exclude strangers from some operations, how are they excluded, how do they know, how clear are the boundaries, how effective is the policing of the boundaries? If the 'strangers' are the general public in one of its guises, has full account been taken of the variability of the general public, or are they being thought of in a stereotyped manner? Will the precautions work for dwarfs, deaf-mutes, illiterates, foreigners, the left-handed?

Within an organization, when hazardous operations take place at particular 'sites', how adequately have these sites been examined for unforeseen problems, bearing in mind the many, many properties of the equipment which is there, including the safety equipment? Do men warm their tea on the tops of the furnaces, hang their coats on the switches? Do they switch the safety equipment off for any reason? What happens if an emergency arises then? Could the safety equipment continue to function in a way which could be inconvenient? Can it be switched off if necessary? When site visits are arranged to agree about aspects of a situation, how clear is it that agreement has been achieved? Can both sides spell out what it is they are agreed on? Is there a check list? Should the check list be extended? Does it cover the behaviour of strangers on the site? For organizations planning any kind of construction, particularly on a large scale, these questions seem to need to be asked for as many as possible of the sites being created. In some ways this might make the task of design on a large scale impossible or, at least, much more difficult, but it should lead to an improvement in the standards of design, insofar as they relate to hazards.

4.6. *Summary*

This chapter has presented the data gathered from an examination of reports into the conditions preceding three major disasters. After reviewing the main events associated with each disaster, the similarities were set out under eight main headings, as follows: rigidities of perception and belief; decoys; organizational exclusivity; information difficulties; 'strangers' and 'sites'; failure to comply with regulations; minimizing emergent danger; and the nature of subsequent recommendations. From these categories, a preliminary set of factors which might be found in pre-disaster conditions was also developed.

The reports presented a variety of somewhat complex material for examination, and in ordering the material in the above form it has already been necessary to condense much of it. But even when this has been done, the categories constitute a rather unsatisfying list, because no clear relationship is apparent between the various elements on this list.

The next stage of the investigation, therefore, is to search for some helpful and acceptable framework which could bring some unity to the analysis without doing too much violence to the data. At the same time, material from other public inquiry reports can be collected and related to that already presented. The following chapters contain accounts of this process, beginning in the next chapter with the presentation of a definition of disaster which seems to be appropriate to present concerns, and which is also compatible with the theory that is later developed.

5. The incubation of disasters

The examination of three disaster reports in the preceding chapter enabled us to identify categories of events associated with the periods of time before these disasters occurred. In order to build upon this analysis, we now have to look at two related problems [1]. We have to consider the possibility that the patterns of events associated with the Aberfan, Summerland and Hixon disasters were so unusual or idiosyncratic that they are of little use when applied to other situations: but if we conclude that these patterns are likely to be of more general applicability, we also have to try to understand why it should be that these kinds of pattern are associated with the pre-conditions of disasters, by finding a more general theoretical explanation which will embrace all of them.

In order to tackle the first question, an additional ten reports of public inquiries into major accidents or disasters in Britain during the period 1965–75 were carefully scrutinized in the same detailed manner that the first three had been [2]. The resulting analyses provided no contradictory cases to suggest that the approach already taken should be revised substantially, but rather reinforced and strengthened the interpretation of the pre-disaster period already presented. Turning, then, to the second question, of trying to identify a way of understanding the relationships between the kinds of event occurring in the pre-disaster period, it seemed to be helpful to locate these events in an appropriate time sequence. Once this had been done, attention could be directed to the question of the knowledge and information available about an impending disaster during the pre-disaster period.

In this chapter, a developmental sequence based upon the analysis of the thirteen disaster reports will be presented but, before this is done, it will be useful to review some of the available definitions of disaster, in order to identify a definition which will be appropriate to the sequence to be presented.

5.1. *Definitions of Disaster* [3]

There is no clear definition of disaster which is wholly appropriate for use in trying to gain an understanding of the events which lead up to such disruptive incidents. As with many words in everyday use, there seems to be no single, precise notion underlying the common usage of the term, waiting to be encapsulated perfectly by means of a few words of definition. For one thing, an elusive line separates the accident from the disaster, a line which can never be finally drawn because any accident may be regarded as a disaster by those close to it. The word is a 'sponge concept', it has been suggested [4], which may be used to refer to the agent causing the events studied, such as a hurricane or a fire; to the physical impact of the agent; to an evaluation of the physical event; or to the social disruption created by the event. Moreover, although the term 'disaster' has advantages over alternatives such as 'calamity', 'catastrophe' and 'cataclysm', some writers [5] have gone so far as to suggest that no absolute definition of the term is possible, since recognition of disaster always takes place in relation to usual prevailing circumstances. Thus, in this view, a disaster, like an epidemic in medical terminology, may be taken to be 'a significant departure from normal experience for a particular time and place'.

Some of the subjective factors which influence the definition of an event as a 'disaster' have even been included in a somewhat ironically formulated 'Law of Inverse Magnitude' [6] which proposes that, in order for an event to be classed as disastrous, the magnitude of the death and destruction associated with it must be increased 'by an undetermined but powerful constant' as physical and emotional distance from the event is increased. And in considering the factors which influence the social definitions of particular events as 'disasters', it would seem to be necessary to add to physical, emotional and cultural distance some account of the 'social loss' incurred, using this notion [7] to cover an amalgam of the prestige of the individuals killed or injured, their position in the society and their expected likely contribution to community affairs. In labelling an event as a disaster, it is also difficult to escape the need to take into account in some way the sense of shock or surprise which an accidental event engenders, a response which itself arises from a complex of factors and issues.

The ideas which are associated with the term 'disaster', then, have an unclear and amorphous quality; as a result, the choice of a definition of the term has generally been bound up with the purposes and interests of the investigator using it. For this reason, the extensive variety of permutations of definitions of disaster which pervade the literature will not be set out here but, instead, one or two more general points will be mentioned in order to enable us to develop a definition of disaster which is relevant to the present investigation.

In one of the most useful recent discussions of disaster phenomena, Western [8] takes a definition of disaster as: 'the relatively sudden and widespread disturbance of the social system and life of a community by some agent or event over which those involved have little or no control', a definition which is appropriate to his concern to establish an epidemiological base for the study of disasters. By contrast, sociologists who have been more concerned with the element of 'perception of disaster' which has already been alluded to, have wanted to include this element in their definitions. Thus, Carr [9] defines a 'catastrophic change' as 'a change in the functional adequacy of certain cultural artefacts', and Killian [10] defines disaster as 'a basic disruption of the social context within which individuals and groups function, or a radical departure from the pattern of normal expectations'.

As we suggested earlier, these writers and others [11] have focused their attention on the post-disaster phase, stressing the extent to which the social structure and functioning of a community are impeded after a disaster, treating any prior events in a cursory manner, if at all. But there is a crucial ambiguity in the kinds of definition which they present, for when a disaster strikes, there are two sources of cultural disruption rather than a single one. The destructive forces unleashed disturb the physical environment and disrupt the everyday conception of the physical world for those affected. But the second type of cultural disruption lies in the occurrence of the event itself, and in its propensity to provoke the question: 'How could such a thing come to happen?' The first kind of disruption has a physical rather than a social cause, and lends itself readily to the 'post-disaster' focus of interest, but the second type of disruption cannot readily be considered within a framework which places time at 'zero' at the moment of impact. This second type is of much more significance than the first in trying to gain an understanding of the nature of disasters, of the reasons why disasters emerge, as opposed to an account of what happens when disasters strike. The issue in this latter case requires a discussion of the manner in which a group or a community is 'surprised', and this question cannot be tackled intelligently without some discussion of the prior state, of the unawareness which made the surprise possible.

With this in mind it seems appropriate to define a more limited kind of disaster, but one which identifies particularly those events which are relevant to the sociological study of the causes and pre-conditions of disaster. Drawing upon the work of other writers, this more limited kind of disaster may be considered as an event, concentrated in time and space, which threatens a society or a relatively self-sufficient subdivision of a society with major unwanted consequences as a result of the collapse of precautions which had hitherto been culturally accepted as adequate [12]. This definition has the advantage of including instances where the amount of physical damage is not great and the number of fatalities is not inordinately high, as a result of chance factors, but where the mishap which has occurred reveals a gap in defences which had hitherto been regarded as secure, so that alarm and the cultural readjustment of expectations follow. The death of two persons from smallpox in London in 1973 is an example of such an event. Public disruption as a consequence of the outbreak was small, but the outbreak arose as a result of a late diagnosis, the disease had been contracted in London, in a medical laboratory, and there had been a failure to trace primary contacts. These features made it necessary to assess the reasons for the collapse of existing safeguards at a full-scale public inquiry. This case is discussed in more detail in Chapter 6.

5.2. A developmental sequence

Having set out the above definition, we can discuss the kind of time sequence that might be appropriate for the consideration of such an event, taking into account not only the three reports on Aberfan, Hixon and Summerland already discussed, but also the ten additional reports from public inquiries during the period 1965–75 mentioned above. When the sequence which has been developed is set out (Table 5.1), it will be possible to use it to guide the investigation of additional public inquiry reports relating to accidents and disasters during that period.

5.2.1. Stage I

The construction of this sequence follows from the essentially sociological definition of disaster as a challenge to existing cultural assumptions set out above. This suggests that the sequence should commence at a starting point where it can be postulated, for the purposes of investigation, that matters are reasonably 'normal'. The set of culturally held beliefs (Stage Ia) about the world and its hazards are at this point sufficiently accurate to enable individuals and groups to survive successfully in the world. Accident analysts make a similar assumption at an individual level when they suggest, for example, that accidents do not occur when a person copes 'with the true situation presented to him' [13]. This level of coping with the world is achieved at the individual and at the group or community levels by adherence to a set of normative prescriptions (Stage Ib) which are consonant with accepted beliefs. These prescriptions about the culturally accepted precautions which are thought to be advisable if hazards are to be avoided with an acceptable level of risk are embodied in laws or codes of practice; or, less formally, they are part of the commonsense set of views of what is safe practice in handling a given situation held by those involved in it: the 'mores' or 'folkways' of safety. When unfortunate consequences follow on a violation of these formal laws or codes, or when accidents occur as a result of departures from precautions prescribed in mores or folkways, there is no need for any cultural readjustment, for indeed, such an occurrence serves to strengthen the force of the existing prescriptions.

In discussing this initial 'normal' stage, it is important to dispel certain 'naive functionalist' implications which arise from the definition given earlier, by pointing out that knowledge about hazards, safety and appropriate precautions is differentially distributed within society. Equally, the ability to take decisions about the level of hazard to which others will be exposed is also differentially distributed. And having made this point, it becomes clear that the answer to the question 'who is surprised when

TABLE 5.1. *The sequence of events associated with the development of a disaster.*

Stage I	*Notionally normal starting points* (a) *Initial culturally accepted beliefs* about the world and its hazards. (b) *Associated precautionary norms* set out in laws,.codes of practice, mores and folkways.
Stage II	'*Incubation period*': the accumulation of an unnoticed set of events which are at odds with the accepted beliefs about hazards and the norms for their avoidance.
Stage III	*Precipitating event:* forces itself to the attention and transforms general perceptions of Stage II.
Stage IV	*Onset:* the immediate consequences of the collapse of cultural precautions become apparent.
Stage V	*Rescue and salvage—first stage adjustment:* the immediate post-collapse situation is recognized in *ad hoc* adjustments which permit the work of rescue and salvage to be started.
Stage VI	*Full cultural readjustment:* an inquiry or assessment is carried out and beliefs and precautionary norms are adjusted to fit the newly gained understanding of the world.

the large-scale accident does occur?' is an important one for the present analysis. In 11 of the 13 reports studied there was complete surprise, in the sense that no-one had predicted or anticipated precisely the kind of accident which did occur. In the case of Aberfan and of the smallpox outbreak, warnings had been given by individuals or by groups concerned, which had not reached, or had not been taken into account by, major institutional bodies concerned. How, for example, a future major earthquake in San Francisco should be classified when it has been so universally predicted remains a moot point.

5.2.2. *Stage II*

A disaster or a cultural collapse occurs because of some inaccuracy or inadequacy in the accepted norms or beliefs but, if the disruption is to be of any consequence, the discrepancy between the way the world is thought to operate and the way it really is rarely develops instantaneously. Instead, there is an accumulation over a period of time of a number of events which are at odds with the picture of the world and its hazards represented by existing norms and beliefs. Within this 'incubation period' a chain of discrepant events, or several chains of discrepant events, develop and accumulate unnoticed. Existing cultural precautions may be thought of as dealing with known and clearly defined hazard problems, but during the incubation period one of the set of vague and unperceived hazard problems which are 'waiting in the wings' begins to be covertly delineated. In fact, more than one unknown danger often begins to accumulate, and a disaster investigation will commonly uncover a number of discrepant events unlinked to the disaster which actually happened.

For discrepant events to build up in this way, it is clear that they must all fall into one of two categories: either the events are not known to anyone; or they are known about but not fully understood by all concerned, so that their full implications are not understood in the way that they will be after the disaster. Developing the categorization from Chapter 4, it may be suggested that in the reports examined the reasons that such events were able to accumulate either unnoticed or not fully appreciated can be summarized as follows:

(1) Events were unnoticed or misunderstood because erroneous assumptions were made. These may have arisen as a result of institutional rigidities of belief and perception, such as the assumption by all those concerned with mining prior to Aberfan that tips were discard, and therefore unimportant. They may have arisen because one problem which was perceived acted as a decoy to draw attention away from another more serious problem; for example, in the Hixon accident,

several individuals concerned checked whether the large transporter carrying a heavy electrical transformer could cross a level crossing without electrical arcing occurring between the transporter and the overhead electric cables, but no-one thought to check whether the long slow vehicle could clear the crossing within the 24-second warning period of the automatic gates. Events were also neglected because complaints of danger from non-experts outside the organization concerned were too readily dismissed as a result of the erroneous assumption that such persons were uninformed alarmists.

(2) Discrepant events were unnoticed or misunderstood as a result of problems in handling information in complex situations. There may have been an excess of information, the crucial messages may have been concealed in a mass of 'noise' [14], or those handling the messages may have been busy or preoccupied with many other matters.

(3) Some events which offered a warning of approaching danger passed unnoticed, or were misunderstood as a result of the common and well-documented human reluctance to fear the worst, which has already been mentioned in Chapter 3 [15], so that danger was frequently belittled even when its onset was noticed.

(4) Finally, where formal precautions were not fully up-to-date, violations of formal rules and regulations came to be accepted as normal.

Lawrence [16], in his model of events leading up to accidents which was discussed in Chapter 2, notes that individual workers may respond in various ways to signs which could alert them to an impending accident: they may fail to perceive the warning event, they may fail to recognize the warning as such, or they may fail to assess the risks adequately. There are clear similarities between these categories and those being set out in the present analysis. However, there is a further category which Lawrence notes in his discussion of pre-accident behaviour which is not very prominent in the 'incubation periods' of the disasters so far analysed. This category is a 'failure to respond effectively to a warning which was recognized and accurately assessed', which accounted for some 13% of the accidents in his survey. This discrepancy between the two models appears to arise because Lawrence's model is intended to deal with accidents which typically result from the occurrences of only one or two errors, and which therefore typically have a short incubation period; whereas the incubation periods for larger disasters seem to be much longer, often extending to several years and including events which are spread over a wide organizational area. Lawrence's model is intentionally restricted to cases in which an error or errors lead to an accident almost immediately, and a failure to take effective action in response to a warning is not uncommon immediately before an accident. However, such a failure can only occur as the last link or almost the last link in the causal chain. If the *only* cause of an incident is an inappropriate response to a recognized warning, the incident is more likely to be one which we characterize as an accident: by contrast, in a pre-disaster situation, given the typically large accumulation of predisposing factors, the nature of the last error is relatively unimportant. Of the thirteen disaster reports examined, two were partially caused by inappropriate responses, and only 2% of the errors contributing to the three major disasters at Aberfan, Hixon and Summerland could be characterized as 'inappropriate response to recognized danger'.

There is always, of course, an infinite network of prior causes which could be traced back for any accident, or indeed any event, but it is important to realize that this is not the incubation network. The incubation network refers only to those chains of events which are discrepant, but are not perceived or are misperceived. It is meaningful to compare accidents and disasters only in terms of incubation networks, and not in terms of sets of infinite causal chains.

In trying to pinpoint the moment in time when a particular incubation period begins for inherently hazardous activities like mining or seafaring, there is a temptation to start at the point at which man first started to dig or to build boats. But given the general recognition of the hazards of such activities, the start of the incubation periods can be located at the point where discrepant events begin to accumulate unnoticed. A crude operational dating of the commencement of an incubation period can be obtained by taking the point at which disaster inquiries start to construct their causal chain. For the 13 reports examined, the incubation periods measured on this basis are as in Table 5.2:

TABLE 5.2. *Lengths of 'incubation period' for 13* disaster reports examined.*

	Incubation period							
	Less than 1 month	1–3 mths	3 yrs	8 yrs	10 yrs	12 yrs	circa 20 yrs	circa 80 yrs
No. of cases	2	2	3	1	1	1	1	1

* One accident had no cause attributed to it, and an incubation period could not therefore be defined.

5.2.3. *Stage III*

The incubation period starts when the first discrepant event occurs unnoticed and it is brought to a conclusion by a *precipitating incident* which produces a transformation, revealing the latent structure of the events of the incubation period. A situation which had been presumed to have one set of properties is now revealed as having different and additional properties which must be interpreted differently. The precipitating event forces itself to the attention because of its immediate characteristics and consequences—a burning building, or an explosion cannot be ignored—and it makes it inevitable that the general perception of all of the discrepant events in the incubation period will be changed.

Firstly, and paradoxically, in terms of the prevailing and most powerful cultural view of the situation up to this point, the incidence of the precipitating event is in a strict sense unpredictable. The event gains part of its force from this very unexpectedness—although it may have been predicted by dissidents or fringe groups of heretics, the precipitating event forces more general recognition.

Secondly, in most cases, this more general recognition of the need for a new interpretation of the situation is forced by the immediate physical properties of the precipitating event, which may be purely technical in nature, as when an explosion or a crash occurs, or a component finally breaks; or may be 'socio-technical', in that a series of human actions may help to create the event as, for example, when an unattended mental patient lights a match in a high-fire-risk hospital ward.

Thirdly, the precipitating event has links with many of the chains of discrepant events in the incubation period: for just as a positive organizational achievement requires a chain of correct acts and decisions if it is to be of any significance, a large disaster requires an extensive chain of errors. A single error leading to a single accident is readily explicable, readily traceable, readily understandable, and fairly easily accommodated into the culture. For a large-scale disruption of cultural expectations it is necessary, as in a good detective novel, to accumulate a sufficient number of unheeded or ambiguous factors to achieve a complete transformation [17].

More immediately, it may be noted that it is possible to illustrate the difference between disasters and accidents already referred to by comparing figures collected by Lawrence [18] for 405 gold-mining accidents with an analysis of three of the larger disaster reports studied (Aberfan, Hixon, Summerland), as set out in Table 5.3.

TABLE 5.3. *Errors and fatalities for some accidents and disasters.*

	No. of accidents	No. of fatalities	No. of 'human errors'	Errors per fatality	Errors per accident
Gold-mining accidents	405	424	794	1·87	1·96
3 major disasters	3	205	191	0·93	Errors for each accident† 36, 61, 50

† A mean for three cases would not be particularly meaningful.

Although the number of errors per fatality in these cases is of a similar order (1·87 as compared with 0·93) the mean number of errors which contributed to each gold-mining accident is 1·96, whereas the accumulation of errors creating the conditions which led to each of the three major disasters were 36, 61 and 50.

In the case where an event which could potentially reveal the nature of the incubating chains of discrepant events is not catastrophic, the forces released are not strong, and the new interpretation is not a compelling one. Thus at Aberfan, the fears of the local council about tip safety were dismissed [19], and the inquiry into the London smallpox outbreak made reference to a laboratory safety expert, Dr Darlow 'crying in the wilderness' when criticizing the laboratory procedures which eventually led to the outbreak [20].

Only when a non-catastrophic realization is achieved by a powerful and prestigious body can a cultural redefinition arise without the impetus of a large-scale physical precipitating event. A review of safety in an hotel which is to be used by visiting royalty may produce results in the improvement of safety which could only otherwise have followed on a major fire.

A final point may be made about the nature of the precipitating event. Carr's long-standing distinction [21] between disasters as 'instantaneous' or 'progressive' has already been noted. In the first case, the precipitating event and its physical consequences follow closely upon one another. In the second case, however, there are two possibilities: a progressive disaster may be produced by a single precipitating event, which is followed by many repercussions. Alternatively, a progressive disaster may result from a series of precipitating events following closely on one another's heels, and producing successive surprises and a need for successive readjustments. 12 of the 13 reports so far examined fell into Carr's 'instantaneous' category, but the remaining one, on the London smallpox outbreak, revealed two precipitating events: first, that the initial undiagnosed illness was smallpox, and second, a fortnight later, that two primary contacts had been missed and were also suffering from smallpox.

5.2.4. *Stage IV*

The precipitating event is followed immediately by the onset of the direct and unanticipated consequences of the failure, an onset which, as Carr has noted, occurs with varying rate and intensity, and over an area of varying scope.

5.2.5. *Stage V*

Closely related is the following stage of rescue and salvage when rapid and *ad hoc* redefinitions are made of the situation by participants in order to permit a recognition of the major features of the failure. In situations of physical disaster, immediate day-to-day modes of accounting for events are often threatened and, in the face of such threats, the victims and onlookers must firstly assure themselves that they are not losing their sanity and that these unprecedented events can be attributed to a fire or an explosion or some similar cause. This may be thought of as a first-stage cultural readjustment to the precipitating event, in which prolonged analyses are not undertaken, but only the minimal recognition of changed circumstances necessary to deal with the immediately pressing problems of rescue and mopping up following the disaster. Some individuals find difficulty, of course, in making this *ad hoc* adjustment for a variety of psychological and emotional reasons [22]. Many of those affected by disaster suffer shock [23], and the after-effects of the experience may persist for years, but from the point of view of the group, it nonetheless makes sense to refer to a period of initial cultural readjustment.

5.2.6. *Stage VI*

When the immediate effects of the onset have subsided, it then becomes possible to carry out a more leisurely and less superficial assessment of the incident, attempting to discover how culturally approved precautions could have turned out to be so inadequate, tracing the now-revealed pattern of events which had developed in the incubation period, and considering the nature of the adjustments which now need to be made to beliefs and assumptions and to laws and statutes.

If the community is to ensure, in the phrase which is often used, that 'this must never happen again', existing ideas about hazards and the means of avoiding them must be reviewed and revised. The new knowledge which makes it possible to carry out these revisions and reinterpretations is the knowledge which was already possessed, though imperfectly, in one form before the incident, but which has now been transformed by the catastrophe into a new configuration which has a different set of meanings and interpretations. A public inquiry or some similar mechanism is commonly involved in this cultural restoration process [24]. Again, as was noted in Stage I, this cultural adjustment is limited by the amount of disagreement which prevails among groups about the effectiveness of any new precautions adopted.

It has already been suggested that the set of events which develop during an incubation period, and which are revealed by the disaster and the inquiries which it provokes, will include some events which are not part of the network leading to the accident, but which form an as yet unlinked branch of some putative future failure of foresight. Many inquiries throw up errors and breaches of good practice, potentially threatening fragments of behaviour, which, if left to themselves, might or might not have led in the future to unfortunate consequences. Thus in the inquiry into the Coldharbour hospital fire [25] the 30 deaths were attributed to fire and fumes resulting from a flame lit in the hospital ward while the ward was unattended. The inquiry also revealed however, as factors *which did not contribute at all* to the particular incident, that there was an ambiguous allocation of responsibility for the locking of the ward doors, and no agreed procedures for unlocking them in the case of fire, that the fire prevention officer had too great a work load, that no clear fire drill had been established, that a fire door had been wedged permanently open and that there was an unacceptable delay in the arrival of the Fire Brigade following the emergency telephone call which was made. Had the fire not occurred when it did, a new scenario could readily be provided for a new and more extensive disaster incorporating some of these additional causal features.

5.3. *An example of the sequence*

The sequence model presented above may be illustrated by reference to one of the reports from the eleven-year period studied. The report describes one of seven major mining accidents which occurred in the period, an accident in which 31 men were killed and one man was seriously injured in an underground explosion. This example has not been chosen as a clear instance of the definition of disaster presented above: the notion of what constitutes a 'disaster' within a hazardous industry like mining, where responses to accidents are highly routinized, may be regarded as open to debate. However, the major disaster reports which

THE INCUBATION OF DISASTERS 93

have been analysed are characterized by a particularly extensive and complex set of events leading up to them, and this major mining accident at Cambrian Colliery, Glamorgan, is more suitable as an illustration of what is meant by an incubation network, because the network here is clearer and less extensive than is the case with the larger disasters studied.

5.3.1. *Stage I*

The notionally normal stage refers in this case to the beliefs held about the hazards of mining and the precautionary procedures needed to deal with them in the area of the Cambrian Colliery, Glamorgan, up to 7 May 1965. In the words of the official report [26]:

> Everyone concerned at the colliery regarded the Pentre Seam and the P26 face in particular as virtually gas-free. The management were no doubt fortified in this belief by the fact that the statutory mine samples never showed more than 0·38 per cent firedamp ...

The pit was thus one in which there was virtually no firedamp, a belief confirmed by regular checks, and therefore a pit in which there was little need for anxiety about ventilation and procedures associated with it.

5.3.2. *Stage II—Incubation period*

One of the properties which is characteristic of the incubation period before a disaster is its ability to be transformed by the precipitating event; to convey this property, the events in the incubation period will be presented firstly, in Figure 5.1, as they were understood up to 1240 on 17 May 1965, and secondly as they came to be appreciated after this point in time, in Figure 5.2. Reviewing the events presented in Figure 5.1, it can be seen that in this 'virtually gas-free' pit, a new face came into production in January 1965. In the process, a temporary air-crossing between two tunnels had been constructed with access holes, rather than a full air lock, to allow men to pass from one tunnel to the other without disturbing the ventilation of the face. The construction of this air-crossing was not of very high quality, but ventilation was not a high priority issue: in any case it was intended shortly to build a permanent air-crossing, and work had started on this, although it had been suspended, because of the low urgency of ventilation, to allow other more important work to go ahead.

In the course of the normal process of mining, an old district (P11) had been contacted on 7 May. There was not thought to be any problem here with methane, and this was confirmed by tests, but the link had to be sealed off anyway, and after six days' work this was completed, methane tests continuing negative. On 17 May a pair of ventilation doors linking with another face were thought to have been left open. There was much activity on the morning of 17 May, owing to breakdowns and repairs to the conveyor and the haulage system. This brought more men than usual to the P26 face, and required them frequently to use the access holes in the temporary air-crossing, a matter which was not particularly remarkable in the absence of ventilation problems. Following the breakdowns of the morning, there was some pressure for production to resume, when at 1215 there was an electrical fault in the plough motor. Electricians were called to repair it and at 1240 they were testing the switch in connection with this repair.

Turning from this manifest pattern of events as it would have been presented to an observer at the pit up to this point in time, the manner in which this is transformed by the addition of the latent events shown in Figure 5.2 may now be considered. Reviewing the situation as the Committee of Inquiry did, after the explosion which occurred at about 1240, it can be seen that the belief that the pit was firedamp-free had been erroneous since 7 May, when the old district had been contacted. As a result, the practices with regard to ventilation are recharacterized as complacent. The poor construction of the air-crossing should not have been tolerated, and the low priority given to the new air-crossing left the face with a poorly constructed air-bridge. The frequent use of access holes on 17 May, which was tolerated since it was believed that there were no ventilation problems, was nonetheless poor pit practice and led to a further crucial reduction in ventilation efficiency.

When the old P11 district had been contacted, methane had been admitted, a danger which had not been recognized, and methane seepage had continued during the period of the closing off of the link, so that firedamp had accumulated unnoticed. The further reduction in ventilation produced by the misuse of the access holes allowed an explosive mixture to build up.

The ventilation doors which had possibly been left open again represented an example of 'poor pit practice', but had no connection with the explosion in this case.

The electricians called in to repair the electrical fault had, for some reason (possibly the pressure to resume production), tested the switch without the gas-tight cover, causing a spark which ignited the firedamp–air mixture.

In this example, the major unnoticed factor is the presence of methane in the pit, undetected because the standard procedures failed to recognize

THE INCUBATION OF DISASTERS 95

conditions in which dangerous concentrations of methane could build up if 'layering' of the gases was accompanied by a reduction in the ventilation flow rate. A knowledge of this factor transforms the interpretation of many of the other events in the network, stressing some of their properties which were not evident before the explosion. As the official report noted, with low firedamp content and excess air in the pit:

> The defects of the air-bridge and the way in which it was used were accepted as being of little consequence, and a sense of complacency with regard to ventilation was engendered in everyone concerned. This sense of complacency continued after the P11 district was contacted, even though the effect of the holing was to change quite drastically the distribution of ventilating pressure [27].

5.3.3. Stage III—Precipitating event

The precipitating factor in this case was evidently the explosion which occurred, bringing to an end the ambiguities of the incubation period with a flame passing 300 metres through the mine, and a blast extending over double this distance [28].

5.3.4. Stage IV—Onset

The immediate consequences of the explosion were apparent in the extensive damage in the mine, in the deaths of 31 of the men working in the area, and in the inflicting of serious injury upon another man [29].

5.3.5. Stage V—Rescue and salvage—First-stage readjustment

Immediately after the explosion, one overman who had been thrown over by the blast went back to the face, meeting the deputy and others, but found it difficult to see in the dust and smoke. They returned to the underground telephone and raised the alarm. The men from another district were called in to help with rescue operations, recovering one injured miner but being unable to proceed further. The Porth Central Mine Rescue Station was alerted at 1305, a rescue brigade arriving at 1317. The first team made one approach to the face and returned as instructed before 1500. Other teams made subsequent rescue sorties, until eventually the atmosphere cleared and the bodies of the victims were recovered [30].

5.3.6. Stage VI—Full cultural readjustment

The major instrument in this final process, particularly in a hazardous industry like mining where there is a degree of routinization in the handling of the after-effects of major accidents of this kind, is the process of statutory inquiry and the publication of a report summarizing the findings of the inquiry. The report in this case draws attention to the various examples of poor pit practice noted, whether or not these contributed to the explosion, and makes adjustments to recommended good practice by suggesting two amendments to the officially approved precautions to be adopted in pits. Firstly it is suggested that continuous tests and monitoring for firedamp should be adopted, and secondly it is recommended that instructions for safer modes of testing electrical equipment underground be produced.

5.4. *Discussion*

The sequence set out above disregards the implicit assumption in much disaster research that disasters spring into existence immediately and in a fully developed condition. By drawing attention to the existence of the 'incubation period' and to the precipitating event which draws it to a close, it becomes possible to consider the properties of these features of the pre-disaster period, and to develop research which will increase knowledge of these features.

This model was used to order much of the remaining research set out in subsequent chapters, but other possibilities for research are also presented by the model. Lawrence [31] draws attention to the fact that, in addition to the 'danger-accident-injury' outcome of accident situations, there are other less noticed outcomes of 'danger–accident–no injury' and 'danger–no accident–no injury', which differ from the accident case only as a result of chance factors. In the accident situation, there seems little to be gained from the investigation of such outcomes; but, in investigating the more complex circumstances associated with disasters, it may be fruitful to look at the circumstances of 'near-miss' disasters as a source of comparative data: hazard and accident analyses in the nuclear engineering industry and in aircraft safety programmes are already making use of such an approach.

5.5. *Summary*

To aid in the discussion of events which precede large-scale accidents and disasters, the development of a definition which stresses surprise and cultural (rather than material) disruption has led to the construction of a phase model of events which includes the following six phases: a notionally normal starting point; an incubation period; terminated by a precipitating event; which leads to the onset of the disaster; rescue and salvage dealing with the immediate problems after the disaster; and a

THE INCUBATION OF DISASTERS 97

final stage of full cultural readjustment to the surprise associated with the precipitating event. This phase model, which has been illustrated by reference to a mining accident, will be developed in the further analysis of material drawn from British public inquiry reports. In the following chapter the model is used to organize some of the various phenomena set out in Chapter 4, and to develop some additional ideas which may help us to understand such events associated with the nature and origins of disaster.

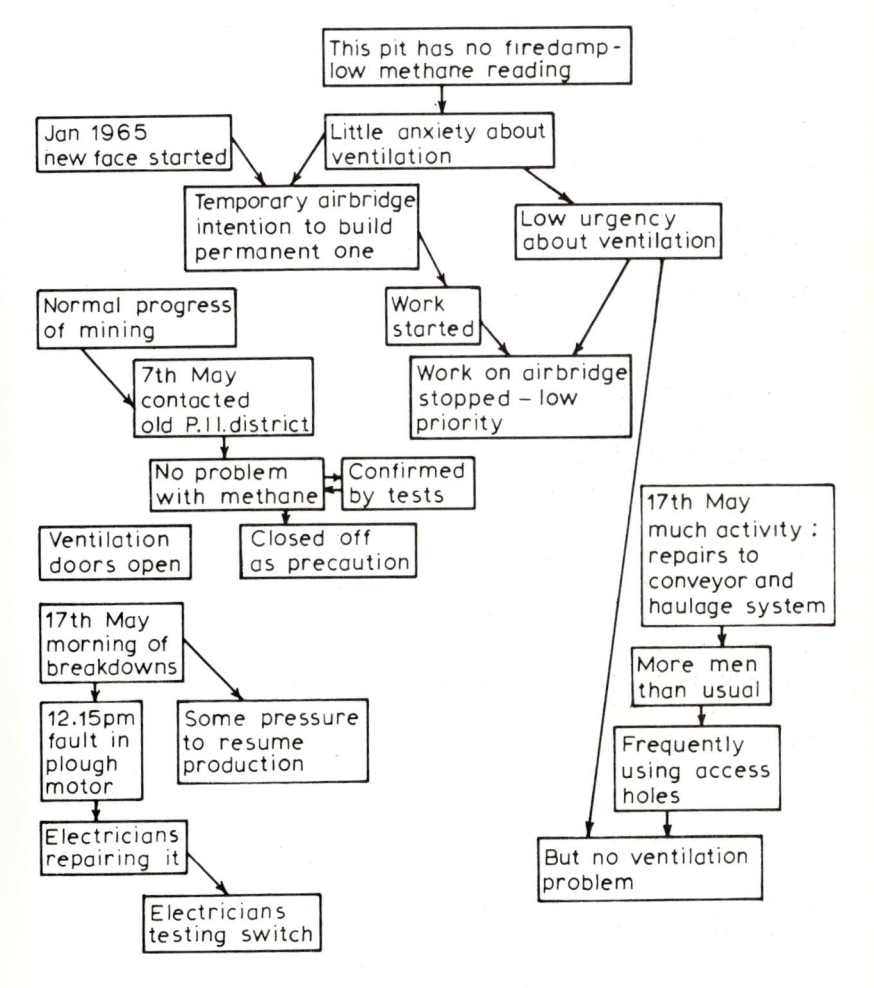

Fig. 5.1. Schematic representation of the situation at Cambrian Colliery immediately before 1240 hours on 17 May 1965.

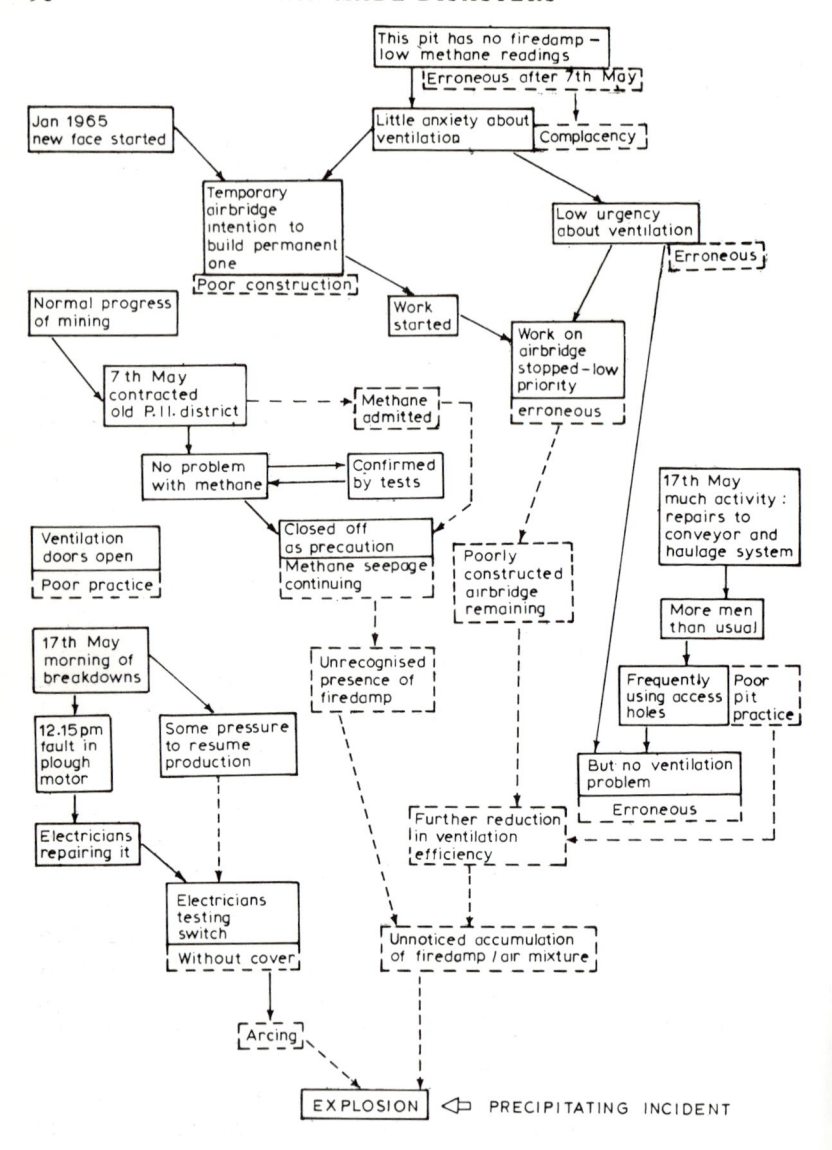

Fig. 5.2. Schematic representation of the situation at Cambrian Colliery immediately after 1240 hours on 17 May 1965.

6. Errors and communication difficulties

Using the theoretical framework developed in the last chapter, we are now better placed to order and understand a number of aspects of disasters, the 10 additional reports examined having strengthened and reinforced the understanding of disaster pre-conditions which was being developed earlier [1]. This framework can be used to sort out the various kinds of event already discussed, and the way in which the kinds of event identified in the analysis of inquiry reports in Chapter 4 fit into the phases of this framework is set out schematically in Table 5.1 [2].

For our purposes the 'incubation period' is the most important phase of the sequence set out in Table 5.1, and most of the events in which we are interested fall into this period (Stage II). Some events leading to failure do occur in the preceding stage, however, and at least two of the public inquiries discussed identify a lack of compliance with existing regulations as contributing to the incidents in question. When such regulations are still regarded as current and applicable, failure to comply with them represents a violation of the existing, accepted precautionary norms, and any accidents resulting from such behaviour do not provoke cultural readjustment. There is no need to adjust existing beliefs about hazards if a major accident takes place, for example, because a company executive fails to observe codes of practice, because a train driver is drunk on duty, or because a pleasure boat is overloaded; although there may still be some need for a review of the manner in which existing precautionary procedures are designed and implemented. Where the regulations violated are known to be inadequate or out-of-date, the situation is somewhat different, and this case is discussed below.

In the present discussion, therefore, those events contributing to disasters which arise from violations of current, culturally accepted regulations are not being regarded as forming part of the 'incubation period'. But most of the events described and investigated in inquiry reports into accidents and disasters (those kinds of event identified in Chapter 4) fall precisely into the 'incubation period' and can be located within Stage II of the phase model. It is within this period that failures of foresight develop, and the kinds of event which may be expected to contribute to the development of disasters during the incubation period may be grouped together into four main types. The events in question may be unnoticed: because of erroneous assumptions on the part of those who might have noticed them; because there were information handling difficulties; because of a cultural lag in precautions; or because those concerned were reluctant to take notice of events which signalled a disastrous outcome. Each of these four types of event may now be reviewed in turn.

6.1. *Events unnoticed or misunderstood because of erroneous assumptions*

Within the incubation period, events which are at odds with existing beliefs begin to accumulate without occasioning comment either because they are not noticed or because their significance is misunderstood. In each of the cases examined, some events contributing to the disaster failed to be appreciated because no-one expected or was alert for such phenomena, or because they were explained away as alternative 'decoy' phenomena, so that their full nature was missed. By its very nature, such a condition is a difficult one to observe without the benefit of hindsight, but clues may perhaps be sought in the manner in which those who dissent from accepted organizational views are treated. If the existing orthodoxy automatically dismisses complaints from outsiders as attempts to claim power by non-expert cranks, the existence of an undue degree of organizational bias and rigidity may be suspected. By the same token, when members of the organization concerned can be seen to be using over-rigid stereotypes of the problems and the people being dealt with, dangerous misperceptions are again likely to develop.

6.2. *Events unnoticed or misunderstood because of difficulties in handling information in complex situations*

Each of the disaster inquiries examined revealed a complex and varied pattern of misunderstandings, ambiguities and failures of communication, some of which contributed to the disaster in question, and others of which were revealed incidentally and found to have no bearing on the accident. Perfect communication will never be possible in any but the simplest of systems, and it may be assumed that many of the misunderstandings and failures to communicate revealed by public inquiries could be readily duplicated in organizational situations where no disastrous outcome occurs. Having said this, it seems reasonable to suggest that there will be some kind of relationship between increasing difficulty in information-handling and increasing likelihood of failures of communication accumulating in such a way as to lead to the incubation of a disaster.

Tasks which must be handled by large organizations will generate a large number of messages within the organization, and are thus more likely to offer opportunity for failures of communication to develop than are tasks which can be handled wholly within a smaller organization. Similarly, where it is felt necessary for a task to be handled by a number of agencies, there is more likelihood of communication failures occurring than when a task can be contained within a single agency. The likelihood of such failures will be increased further by the fact that each organizational unit or sub-unit will have developed its own distinctive sub-culture and its own version of rationality, which may give rise to erroneous assumptions about the portion of the problem which is being handled by other units. Again, the more prolonged, the more complex, the more vague, the more hasty and the more large-scale the task, the more likelihood there is of information-handling difficulties arising. Prolonged tasks are likely to be associated with changes in goals, changes in responsibilities, and changes in administrative roles which create difficulties; large and complex tasks generate more information, vague tasks generate more ambiguities and more distractions, and rushed tasks lead to the overlooking of information because of pressure of work. Two factors which were identified as particularly contributing to the complexity and unpredictability of organizational tasks are the design of large, complex sites, and the management of groups of 'strangers' who have access to such sites.

Since a state of variable disjunction of information is by definition one in which the resources available to handle information are inadequate, such a condition would also be expected to increase the propensity for information difficulties to accumulate in a hazardous manner.

6.3. *Effective violations of precautions passing unnoticed because of cultural lag in existing precautions*

A simple failure to observe existing regulations has been mentioned already, above, but a more complex situation arises when existing precautionary regulations are discredited, but not yet changed, because they are out-of-date or inapplicable to the case in hand. In such cases, as when existing Theatre Regulations were deemed to be inapplicable to the Summerland leisure centre, there may be difficulty in finding an appropriate standard by which to judge the *ad hoc* solutions arrived at, without the benefit of guidance from well-considered formal precautions.

6.4. *Events unnoticed or misunderstood because of a reluctance to fear the worst outcome*

This tendency was a particularly noticeable one in the evidence submitted to the inquiries studied, and it has been noted also by other observers. Clearly, when existing danger signs are not perceived, are given low priority, are treated as ambiguous or as sources of disagreement, and when they are treated as insignificant for psychological or other reasons, another avenue is provided for the accumulation of events which may combine to lead to disaster.

Stage III–V in Table 6.1 cover the area of most pre-existing disaster studies, and no attempt was made in the present study to categorize events falling in these three stages. A final note may be made, however, in Stage VI, of the role played by each of the committees or tribunals of inquiry in *establishing a new level of precautions, and new expectations about their efficacy*. In each case, the report weighed and evaluated the behaviour revealed by its inquiries, marking it down as responsible or irresponsible, as good or bad practice, even when the behaviour being considered may have had no direct bearing upon the particular incident under investigation. Each report then concluded by making recommendations which sought to amend existing beliefs and assumptions, and to establish new norms of behaviour, in order to prevent a recurrence of any similar incident [3].

TABLE 6.1. *Common features observed in the development of three major disasters and their relation to various stages of development.*

Stage of development	Feature	Comments
Stage I Initial beliefs and norms	6. Failure to comply with existing regulations	*Violation of existing precautions*
Stage II 'Incubation period'	1. Rigidities of belief and perception 2. Decoy phenomena 3. Disregard of complaints from outsiders	A. *Events unnoticed or misunderstood because of erroneous assumptions*

Table 6.1 (*continued*)

	4. Information difficulties and 'noise' 5. The involvement of 'strangers'	B. *Events unnoticed or misunderstood because of difficulties of handling information in complex situations*
	6. Failure to comply with discredited or out-of-date regulations	C. *Effective violation of precautions passing unnoticed because of 'cultural lag' in formal precautions*
	7. Minimizing of emergent danger	D. *Events unnoticed or misunderstood because of a reluctance to fear the worst outcome*
Stage III Precipitating event	—	
Stage IV Onset	—	
Stage V Rescue and Salvage	—	
Stage VI Full cultural readjustment	8. Definition of new well structured problems and appropriate precautions in inquiries following the disaster	*The establishment of a new level of precautions and expectations*

TABLE 6.2. *Some examples of the features listed in Table 6.1.*

Feature	Example
IIA	
1. *Rigidities of belief and perception*	Major institutional neglect of tips as a potential source of danger (Aberfan).
2. *Decoy phenomena*	Local residents mistakenly thought that the danger from tips at Aberfan was associated with the tipping of very fine

Table 6.2 (*continued*)		waste, and they withdrew some of their complaints when it was agreed that this would not be tipped.
		Concern of police and transporter crew at Hixon with the danger of arcing on to overhead wires, and not with collision.
3. *Disregard of complaints from outsiders*		Complaints from Aberfan residents not adequately dealt with by the National Coal Board.
		High-handed response from British Rail to haulage company over stalled lorry on crossing prior to Hixon accident.

IIB

4. *Information difficulties and 'noise'*		Poor communication between individuals because of poor personal relations (Aberfan).
		Ambiguous orders: does 'facing traffic' mean facing approaching traffic or stationary traffic at the crossing (Hixon).
		Information buried in a mass of irrelevant material (Hixon).
		Information neglected because of pressure of work (Summerland).
		Police expected government documents to be 'interpreted' for them, but this was not done with the automatic crossings instructions.
5. *The involvement of 'strangers'*		Road traffic using a rail crossing at Hixon were 'strangers' to the rail system.
		Public using the leisure centre at Summerland were 'strangers' with regard to the organizations operating the centre.

IIC

6. *Failure to comply with discredited or out-of-date regulations*		Uncertainty about how traditional Theatre Regulations should apply to a 'new concept' leisure centre such as Summerland.

IID

7. *Minimizing of emergent danger*		Early movements of the tip at Aberfan were not conceived of as leading to major hazard.
		Minor fire at Summerland was dealt with by staff, and there was delay in summoning the Fire Brigade.

The material set out above could be reformulated in an even more condensed form in the proposition that disaster-provoking events tend to accumulate because they have been overlooked or misinterpreted as a result of false assumptions, poor communications, cultural lag and a misplaced optimism. At this level of generality, such a proposition may not be particularly unexpected, but it should be considered in the light of two points which have not so far been strongly emphasized. Firstly, disasters other than those arising from natural forces are not created overnight. It is rare that an individual, by virtue of a single error, can create a disastrous outcome in an area formerly believed to be relatively secure. To achieve such a transformation he needs the unwitting assistance offered by access to the resources and resource flows [4] of large organizations, and he needs time. As Table 5.2 shows, the three major accidents with which this analysis began had been accumulating for a number of years. Small-scale failures can be produced very rapidly, but large-scale failures can only be produced if time and resources are devoted to them.

Secondly, the process of condensation which has been pursued since Chapter 4 has not been merely lumping together many subtly different modes of miscommunication under a single heading. Each of the disasters studied arose, not because of a single factor, but because of the accumulation of complex chains made up of mixtures of individual and organizational events, more complex versions of the kinds of event represented schematically in the diagrams in Chapter 5. When such diagrams are constructed for disasters, the events which are included in them may be events selected from clusters of cultural, institutional, informational, psychological and task-related phenomena of the kind set out in Chapter 4 in the analysis of Aberfan, Hixon and Summerland. The above scheme, therefore, does not simply say that disasters arise because of miscommunication, but suggests that some degree of order may be brought to the infinite array of possible factors which may contribute to the incubation of disasters by classifying them in the above manner. Given the emphasis upon cultural disruption in the present study, it is not surprising that the central organizing principle of this classification should be concerned with information and communication.

6.5. *Boundaries and communication networks*

There is an additional aspect of this concern with communication in the pre-disaster period which was thrown up by some of the cases examined, and which has not so far been discussed. The analysis presented above makes the rather over-simple assumption that information is either available or not available, that knowledge of immediate hazards, or factors leading to them, is either present or absent. This of course is not true, for it is often the case that the information which could prevent a disaster is available to someone, but that that individual may not realize the significance of the knowledge that he has, or he may not be able to pass this information on in time, either because he does not know precisely where it is needed, or because the constraints of habit, lack of authority or lack of resources stop him from passing on his knowledge. In studying the origins of disasters, therefore, it is important to pay attention, not just to the aggregate amount of information which is available before a disaster, but also to the distribution of this information, to the structures and the communication networks within which it is located, and to the nature and extent of the boundaries which impede the flow of this information. Of particular interest are those boundaries which, by inhibiting the flow of information, may permit disasters to occur. This topic will be examined here through a discussion of seven illustrative cases, and the links between these information patterns and the phase classification set out in the preceding section will also be outlined.

Research workers have not totally neglected the study of the problems of communication in the periods prior to disasters; but, when they have examined such problems, their main concern has been: to assess the difficulties faced by public authorities in deciding when to give warnings; to consider whether the information the authorities have is such as to lead to the implementation of emergency plans at appropriate times; and to try to determine the most effective way in which a threatened population can be warned about impending danger [5]. Instead of concentrating almost exclusively upon the manner in which the damage and injury caused by a disaster can be minimized by changes in information-handling procedures, it seems desirable to develop a means of placing the problems of perception and definition in a pre-disaster situation into the wider context of the more general problems of handling information prior to the realization that an emergency situation has arisen. It was suggested earlier that the occurrence of a disaster moves the vague, ill-defined, ill-structured problem which had previously only been dimly perceived to a 'post-disclosure' condition, where it is reformulated as a new, potentially well-structured problem. For this transition to occur, for the incubation period of an emergency to terminate and for the emergency proper to begin, there must be a transfer of information which will change the perception of the problem to a more structured and recognizable form. In one sense, of course, this transformation is one of degree, for a disaster occurs when existing well-structured views of the world fail to take account of all of the relevant factors operating, and while the disastrous event provides the additional unwelcome information necessary to start restructuring the problem, the new understanding itself can only be partial; thus it, too, may be overturned at some future time, whether or not we are aware of its partial nature.

In spite of this qualification, however, it is still worth paying attention to the kinds of information made available as a result of a large-scale accidental event. In most of the cases from which the framework already set out was developed, this information flow tended to occur only as a result of a final, violent precipitating event, where the form of the hazard was a novel or unprecedented one, or where some of the contributory factors had not been encountered before.

But there are other cases where the hazard which emerges is one which is recognized, and has been encountered and handled satisfactorily before and where a large number of warning signs are potentially available, so that there is neither a high degree of novel hazard, nor a large number of wholly concealed 'clues' about hazard in the pre-disaster situation. In these cases, the 'ill-structured' elements of the problem relate merely to matters of the timing and location of a particular hazard, and the realization that it is this particular form of hazard with which one should currently be concerned. If these elements can be defined before the hazard begins to provoke precipitating incidents, it may be possible to avoid an emergency, or to limit the scale of its consequences. The cases discussed below are cases in which the information needed to develop a full awareness of the problem is dispersed among a number of parties who often have different institutional or organizational affiliations. It therefore seems to be worthwhile setting out the patterns of distribution of information in order to try to develop an understanding of the manner in which these particular types of failure of foresight develop [6].

6.6. *Some Cases of Disclosure*

In this discussion, the patterns of information distribution associated with seven incidents will be outlined, and some of the general features of these cases presented in a manner relevant to the previous theoretical framework. The seven cases used were selected from those discussed in 84 official British government reports published during the period 1965–75 [7]. They were chosen because they could be arranged in a series indicating the way in which the theoretical framework presented so far could be further developed and extended in order to take into consideration some of the features of communication associated with the disclosure pattern characteristic of accidents and disasters.

6.6.1. Case I—Use of contaminated fluids at Devonport section of Plymouth General Hospital [8]

In this incident, a batch of dextrose infusion fluid manufactured by Evans Medical Company at Speke in 1971 was not satisfactorily sterilized, and an indication that a satisfactory sterilizing temperature had not been reached was ignored. Part of the batch was sent out to a distribution company at Paignton, and subsequently delivered to the hospital at Devonport, being taken into use in February 1972. In the words of the official report, the following events then took place:

> As a result of a succession of untoward reactions in patients at the hospital on 1st, 2nd and 3rd March, bottles of the sub-batch came under suspicion, and were subjected to bacteriological examination, which confirmed on 4th March that they were contaminated.

The hospital authorities, Evans Medical and the Department of Health in London were informed, and prompt action was then taken to prevent further use of the remainder of the batch.

Trying to summarize the events in the report in order to illuminate the process of disclosure, it can be said that as a result, according to the report, of poor communications and reliance on inadequate procedures within Evans Medical, an undisclosed condition which constituted a potential hazard arose in April 1971. Once the one possible clue to this condition, a temperature recording chart showing insufficient sterilization, had been ignored, there were no other warning signs to personnel in Evans Medical which could possibly have indicated the potential hazard. Consequently, there was nothing to stop the normal processes of organization and distribution used for the dissemination of safe medical supplies also being used to transmit this hazard to several locations remote from its point of generation [9]. Since the bottles remained sealed and boxed, the role of the distributors at Paignton was, of course, neutral with regard to this hazard, as were the actions of the store-keepers and the pharmacist at the hospital. Indications of the hazard began to emerge on 1 March 1972 when a staff-nurse rejected a bottle with small particles floating in it. On the same day, and the two following days 'a succession of untoward reactions' in patients of Mr William Gall, a consultant surgeon, and his colleague Mr Michael Reilly, aroused their suspicions, and they asked the Director of the local Public Health Laboratory to investigate, at the same time suspending further use of the fluid. On 4 March, the suspicion of contamination was confirmed, and the disclosure was complete, enabling the Department of Health and other authorities

ERRORS AND COMMUNICATION DIFFICULTIES

to take routine action to deal with the well-structured problem now revealed.

The relevant zones of interaction relating to this fairly simple and straightforward disclosure are three:

(a) internal interactions within Evans Medical which took place between those responsible for manufacturing and supervising the manufacture of medical supplies, interactions which enabled the contamination to take place, in spite of inspection of the company by the Department of Health;

(b) interactions associated with the distribution of the fluid, which disseminated the hazard, but in other respects remained neutral towards it;

(c) interactions between the medical and nursing staff using the contaminated fluid. Within this context, the suspicion of hazard arose, to be rapidly confirmed as a result of communication with a colleague in an associated institution.

These zones are set out schematically in Figure 6.1.

The process of disclosure was thus concentrated at Devonport, and was a fairly simple one of warning signs generating increasing suspicion of contamination, and action being taken to confirm this suspicion.

To summarize the relevant points of interest arising from this case, it may be noted that (i) the hazard emerged, as is often the case in modern society, at a point remote from the area in which it was generated, and (ii) the hazard could have been disclosed at two points: at the point of manufacture, within Evans Medical, or at the point of use. At both of these points, the systems of interaction within which disclosure might have occurred were localized and closely tied to one particular institution. When disclosure did arise at Devonport, the interactions which led to suspicions being aroused were all located within the hospital, and for confirmation it was only necessary to make use of a routine, established channel of communication with an associated local institution.

6.6.2. *Case II—London smallpox outbreak, 1973* [10]

By contrast with Case I, in which the process of disclosure was relatively simple, localized and confined to a single institution, this case presents a much more complex picture, but one which it is possible to analyse by using and extending the terms already employed to discuss the first case. The events associated with the London smallpox outbreak are discussed at some length here, partly because they are interesting in themselves, but also because they present us with an instance of a situation in which a

(a) Interactions associated with emergence of hazard

[Inspection role of Department of Health]
↕
[Internal information flows in Evans Medical surrounding the emergence of the hazard]

(b) Interactions associated with distribution

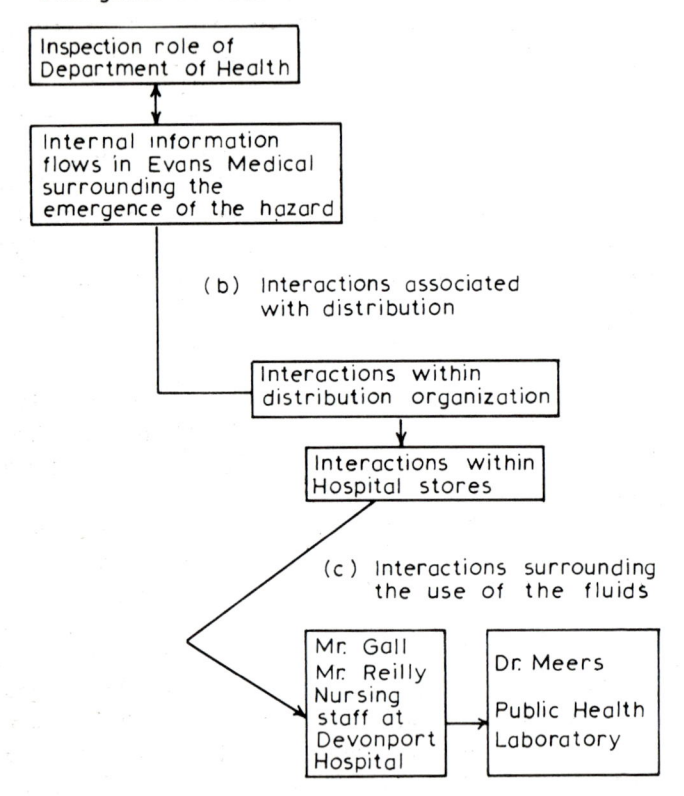

[Interactions within distribution organization]
↓
[Interactions within Hospital stores]

(c) Interactions surrounding the use of the fluids

[Mr. Gall / Mr. Reilly / Nursing staff at Devonport Hospital] → [Dr. Meers / Public Health Laboratory]

Fig. 6.1. Schematic representation of zones of relevant communication and interaction relating to the use of contaminated fluids at Devonport section of Plymouth General Hospital, 1972.

Derived from Cmnd 4470. See note 8.

ERRORS AND COMMUNICATION DIFFICULTIES 111

potentially much more damaging outcome was prevented from occurring as a result of an unorthodox and non-routine communication which could not possibly have been predicted in advance. And, what is even more curious, this pattern of worse disaster being averted as a result of an unpredictable information flow is repeated in each of the two stages of the outbreak.

The smallpox hazard arose on 28 February 1973 when Miss Algeo, a newly appointed laboratory technician in the Mycological Reference Laboratory, observed another technician, Mr Bruno, harvesting smallpox virus in an adjacent laboratory which was part of the London School of Hygiene and Tropical Medicine. During the course of her work Miss Algeo regularly visited this laboratory, which was in the same building, in order to use equipment there. On 11 March Miss Algeo fell ill, and her doctor transferred her to hospital on 16 March with suspected meningitis. Disclosure of the fact that she had smallpox, and thus constituted a public danger while kept in a public ward, did not occur until 23 March. Two untraced contacts contracted smallpox and died a short while later.

Miss Algeo's infection did not manifest itself in a straightforward fashion, possibly as a result of earlier vaccinations she had had, and in any case the medical staff who saw her had no reason to suspect smallpox as a possibility.

Thus, while the condition for disclosure in Case I could be characterized as being concerned with a simple 'contamination/no contamination' possibility, in this case the 'smallpox/not smallpox' question did not arise until relatively late in the process, and disclosure was instead a matter of assembling a number of facts which, together, prompted the realization that it was highly likely that Miss Algeo had smallpox. These items of information are summarized in Table 6.3.

The degree of informedness of the various individuals before 21 March 1973 is summarized in Figure 6.2, which makes it clear that, in general terms, those who were in those sections of institutions where they were likely to know that Miss Algeo had been exposed to smallpox virus did not know that she was seriously ill, while those who knew that she was seriously ill (her superior, Dr Mackenzie, her G.P., and the hospital doctors) did not know that she had been exposed to smallpox. Miss Algeo herself remained unaware of the danger to which she had been exposed.

The situation summarized in Figure 6.2 is thus one in which a number of items of information or clues as to the hazard are available, but they are distributed among many parties and individuals with a variety of different patterns of association. When such a condition arises, there are

TABLE 6.3. *Items of information to be assembled in order to disclose that Miss Algeo was suffering from smallpox.*

Information about general background matters	Information about specific events in February and March 1973
A. That smallpox virus was used in the pox virus laboratory.	
	B. That Miss Algeo went into the pox virus laboratory in the course of her work.
C. That unvaccinated people should not be allowed to enter the pox virus laboratory.	
D.1. That vaccination in general was encouraged in the laboratory, though not necessarily related to smallpox.	
D.2. That smallpox vaccination specifically was needed.	
	E. That Miss Algeo watched virus being harvested on 28 February.
	F. That Miss Algeo was ill.
	G. That Miss Algeo was in hospital.
	H. That Miss Algeo was infected with a pox virus.
I. That watching the harvesting of smallpox virus is dangerous.	

a number of possible resolutions: the hazard may remain unnoticed until it is realized, in an unequivocal manner, as a result of a 'precipitating incident'—further cases of smallpox among primary contacts would have produced this resolution. Or the hazard may be disclosed by the assembly of the relevant pieces of information as a result of com-

ERRORS AND COMMUNICATION DIFFICULTIES 113

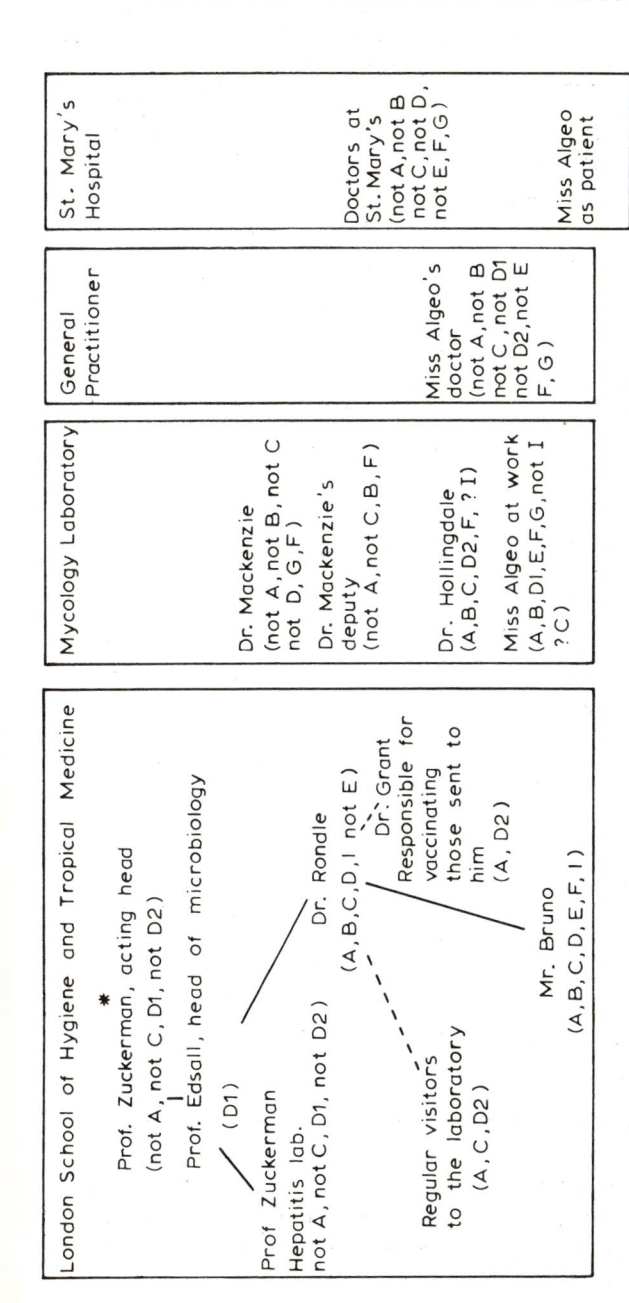

London School of Hygiene and Tropical Medicine	Mycology Laboratory	General Practitioner	St. Mary's Hospital
Prof. Zuckerman, acting head (not A, not C, D1, not D2) * ⎯⎯ Prof. Edsall, head of microbiology (D1) Prof Zuckerman Hepatitis lab. not A, not C, D1, not D2 ⎯⎯ Dr. Rondle (A, B, C, D, I not E) Dr. Grant Responsible for vaccinating those sent to him (A, D2) Regular visitors to the laboratory (A, C, D2) Mr. Bruno (A, B, C, D, E, F, I)	Dr. Mackenzie (not A, not B, not C not D, G, F) Dr. Mackenzie's deputy (not A, not C, B, F) Dr. Hollingdale (A, B, C, D2, F, ? I) Miss Algeo at work (A, B, DI, E, F, G, not I ? C)	Miss Algeo's doctor (not A, not B not C, not D1 not D2, not E F, G)	Doctors at St. Mary's (not A, not B not C, not D, not E, F, G) Miss Algeo as patient

* Professor Zuckerman occupied two institutional roles at this time.   Derived from Cmnd 5626. See note 10.

Fig. 6.2. Distribution of information relating to Miss Algeo's condition before 21 March 1973, arranged according to zones of likely interaction.

For key to information, see Table 6.3.

munications along routine channels. Where such routinized communication channels do not exist, as in Figure 6.2, the only manner in which the further incubation of the hazard, until it becomes a public emergency, can be avoided is by the occurrence of non-routine or even fortuituous communications using non-institutionalized channels.

This, in fact, was the course of events here, for the disclosure process was completed by the five communications summarized in Figure 6.3, three of them crossing organizational boundaries, and one of these being highly unorthodox. Miss Algeo's superior officer Dr Mackenzie initiated all of these cross-institutional communications, impelled by a concern for her condition as possibly having arisen from materials in his own mycological laboratory, and by a puzzlement at certain features of her condition. Dr Mackenzie's discovery (from a skin sample taken during visiting hours!), that Miss Algeo was infected with a pox virus, provided a piece of information which aroused the suspicions of Dr Rondle from the London School of Hygiene and Tropical Medicine, and made him receptive to the final fragment which completed the disclosure of the hazard, when Mr Bruno informed him that Miss Algeo had watched the harvesting of a batch of smallpox virus.

In Case II, the relevant zones of interaction concerned with the hazard created by Miss Algeo's condition, and its eventual disclosure, are five, summarized in Figure 6.4.

(*a*) Internal interactions and information flows within the London School of Hygiene and Tropical Medicine.
(*b*) Internal interactions within the Mycological Reference Laboratory.
(*c*) Informal patterns of association based upon frequent association, and the use of a common 'territory', the pox virus laboratory, by some members of both the London School of Hygiene and Tropical Medicine, and the Mycological Reference Laboratory, including Miss Algeo. This pattern clearly overlaps with (*a*) and (*b*).
(*d*) Miss Algeo's consultation with her doctor.
(*e*) Interactions between Miss Algeo and the medical staff treating her at St Mary's Hospital.

Case II may be summarized in the following terms. The hazard again is one which displayed its effects at a point remote from that at which it arose; in this case, though, the information which could have led to disclosure before the full force of the hazard was felt was not concentrated within a single institution, but scattered among three institutions. Routine interactions and communications were concentrated within these institutions, and also within an area of common territory and routinized

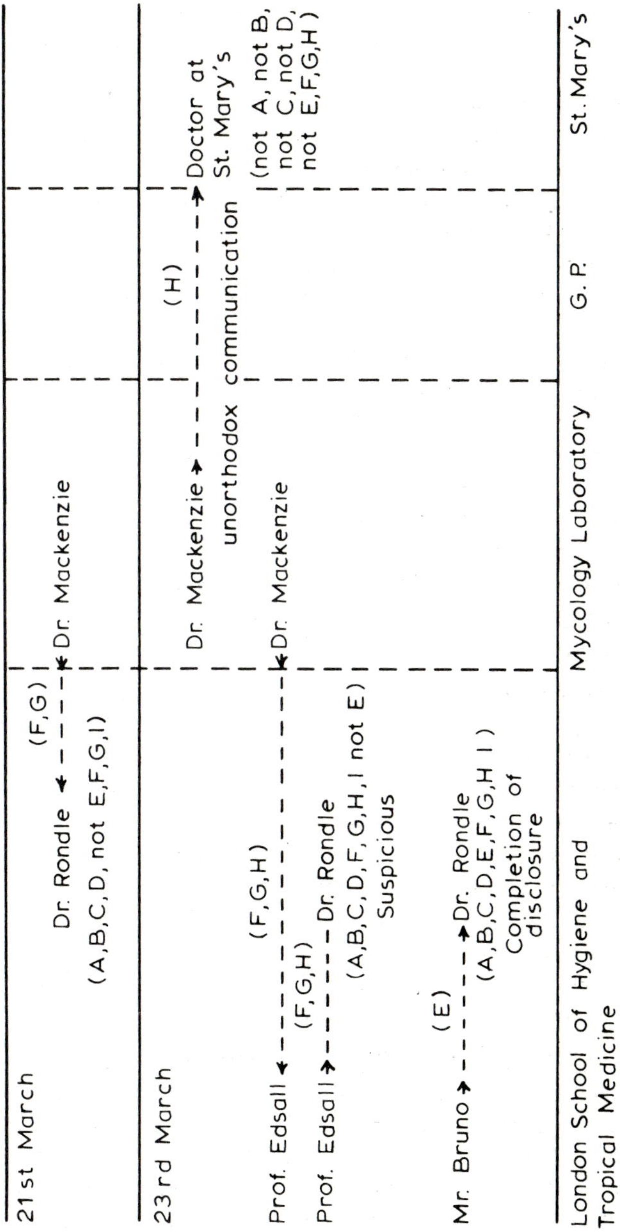

ERRORS AND COMMUNICATION DIFFICULTIES

21st March

Dr. Rondle (F,G) → Dr. Mackenzie
(A,B,C,D, not E,F,G,I)

23rd March

Dr. Mackenzie → (H) - - unorthodox communication → Doctor at St. Mary's
(not A, not B, not C, not D, not E,F,G,H)

← Dr. Mackenzie

Prof. Edsall ← - - (F,G,H) - - - Dr. Rondle
(F,G,H)
Prof. Edsall → - - - - - Dr. Rondle
(A,B,C,D,F,G,H,I not E) Suspicious

Mr. Bruno → - - (E) - - → Dr. Rondle
(A,B,C,D,E,F,G,H I) Completion of disclosure

| London School of Hygiene and Tropical Medicine | Mycology Laboratory | G.P. | St. Mary's |

Derived from Cmnd 5626. See note 10.

Fig. 6.3. Communications taking place along various channels, leading from the information distribution indicated in Figure 6.2 to the disclosure of Miss Algeo's condition 21–23 March 1973.

(a) Interactions within the London School of Hygiene and Tropical Medicine

(b) Interactions within the Mycological Reference Laboratory

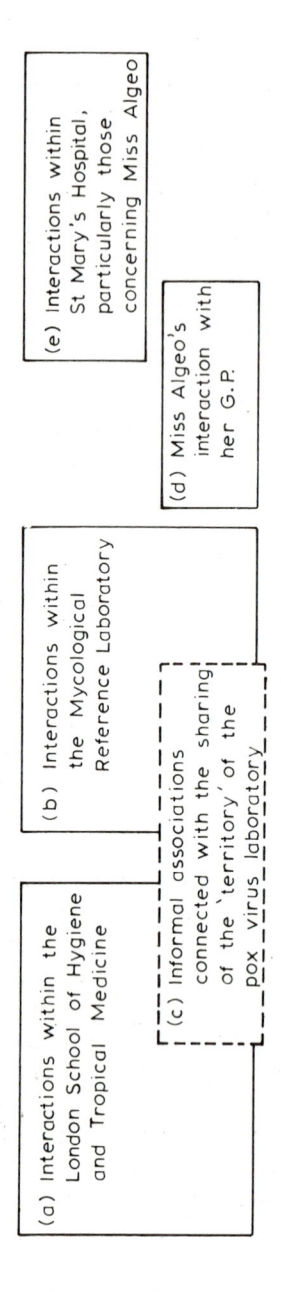

(c) Informal associations connected with the sharing of the 'territory' of the pox virus laboratory

(d) Miss Algeo's interaction with her G.P.

(e) Interactions within St Mary's Hospital, particularly those concerning Miss Algeo

Fig. 6.4. Summary of main zones of interaction in Case II.

inter-organizational communication centred upon the pox virus laboratory. In this situation, the only manner in which disclosure could occur was if the required information flows took place along non-routine channels, and this in fact was what happened.

Curiously and tragically, a similar pattern was repeated at a later stage in the same outbreak, with regard to those primary contacts of Miss Algeo who contracted smallpox. These two people visited a patient in the next bed to Miss Algeo, but were missed in the routine procedures for tracing contacts. They, too, were admitted to hospital as suffering from suspected food poisoning, and remained in public wards until an unusual communication, resulting from a remarkable display of deduction and initative by a social worker dealing with a pension query from Miss Algeo's former neighbour in the hospital, was made. Piecing together a rather confused set of information and linking this with a small Press item about smallpox, the social worker telephoned the hospital concerned and informed them that they probably had two smallpox suspects under treatment. Thus, although two deaths unfortunately resulted, a much larger smallpox outbreak in London was averted by means of unusual and non-routine messages which made for an early disclosure of the hazard.

6.6.3. *Case III—Dudgeon's Wharf explosion*

The third illustrative case which may be mentioned briefly in this section is that of an explosion which occurred at Dudgeon's Wharf, Isle of Dogs, in 1967 [11] when employees of a demolition firm were dismantling oil storage vessels. The owners of the tanks were under the impression that a large demolition company with appropriate experience was at work, and the Fire Brigade, whose representatives visited the site several times during the course of the work, were alarmed at the manner in which the demolition was being carried out, but believed the contractors when they assured the fire officers that they intended to implement the numerous precautions which the Fire Brigade had suggested to them.

In fact, the contractors were a small, *ad hoc* organization and, in their handling of their relationships with the other parties concerned, they appear to have taken few steps to divest the other parties of any misconceptions they may have had about their own status or actions, while remaining relatively assured and confident themselves that they would readily handle the dangers of the task which they were tackling. The problematic feature of this disclosure case is that there seems, at least as far as can be judged from the inquiry report, to have been an element of volition inhibiting the flow of the relevant information about emerging danger between the three main institutional groups concerned.

6.6.4. *Case IV—Loss of m.v. Nicolaw*

The voluntary inhibition of communication about possible dangers, or about conditions which may allow danger to develop, is seen more clearly in the fourth case, drawn from the report of a Court of Inquiry into a Marine Wreck, the loss of the motor vessel *Nicolaw* in November 1969 [12]. In this case it is made very clear in the report that events which were known by some of the individuals involved could have provided forewarning of a possible disastrous outcome, but that these events were effectively concealed as a result of intentional actions on the part of some of those involved.

In the lengthy report on m.v. *Nicolaw*, the events leading up to the loss of this ship, which had been bought for conversion to a dredger, are set out. The Managing Director of the small subsidiary company set up to buy the vessel was given the chance to acquire a large share of the early profits accruing from the initial operation of this vessel; possibly because of this, he concealed aspects of the unseaworthiness of the vessel, passed himself off as a qualified Master, and generally flouted a number of safety regulations. Amazingly, he succeeded in concealing the true situation from a whole range of authorities who had statutory duties with regard to the safety of ships. Described by the report of the Court of Inquiry as 'an ambitious and persuasive imposter', the Managing Director, acting as Master of the ship, died when the ship went down, after giving his life jacket to the second mate, who did not have one.

This tragic case clearly illustrates a particular kind of information difficulty which may be associated with the pre-disaster period. Information difficulties in this class arise during the incubation period from the active concealment of certain aspects of the situation, not with an intent to cause disaster, but in the belief that accidents will not happen. As a result of this concealment, features which might cause alarm to others concerned, particularly the authorities administering safety regulations, are not evident to everyone involved.

6.6.5. *Case V—The two-party pattern*

As a final set of cases relating to communication patterns and disclosure, we shall discuss three marine collisions. In none of these cases was there injury or loss of life, although the material damage caused as a result of the collisions was of varying severity. The first collision [13] was that between two motor tankers, *Esso Ipswich* and *Efthycosta II*, in April 1970 as both were in the charge of pilots approaching Cardiff. The Court of Inquiry found that the collision occurred partly by default of the pilot of *Esso Ipswich*, and partly by default of the Master of *Esso Ipswich*.

Criticisms were made of those in charge of the other vessel also, but these faults were not considered 'causative' in this collision. The collision may be said to have occurred as a result of a combination of miscalculation and poor practice as two vessels approached a slightly unpredictable area of turbulence at the same time.

The second collision [14] took place in November 1971 between motor vessel *Redthorn* and motor vessel *Efpha* in fog off the Sussex coast. The owners of *Efpha*, a large Liberian tanker, gave no assistance to the Court of Inquiry, but the vessel seems to have been proceeding in fog at full speed, probably without even seeing *Redthorn*, and in fact altering course towards the smaller coaster a couple of minutes before the collision. The Captain of *Redthorn* was acknowledged to have been placed in a difficult position, but it was nonetheless felt by the Court that he should not have entered a fog bank at speed, without fog signals, and with an over-reliance on his radar.

The third collision [15] occurred in December 1973 between motor vessel *British Fern*, a large oil tanker, and motor vessel *Teviot*, a 700-tonne refrigerated liquid-gas carrier. The Master and officers of *British Fern* observed the approach of *Teviot* from its first radar sighting, $14\frac{1}{2}$ km away, and then by visual sighting at 3–5 kilometres' distance. As the ships came closer, *British Fern* signalled first of all with flashes of a signalling lamp, and then with siren blasts. All in all, they took what evasive action they felt to be appropriate, but the efficacy of their actions for the avoidance of a collision relied partly upon the Master of the other vessel observing them, and behaving accordingly.

What those in charge of *British Fern* did not know was that effectively, no lookout was being kept on the other vessel. The Captain of *Teviot* left the wheelhouse and lookout position unattended for a period, while he went to the chart-room. He was still in the chart-room when he heard a whistle, and spotted the other vessel less than 50 metres away. Though he then ran to the wheelhouse, he was unable to take any other action before the two vessels collided [16].

Although these collisions are not to be regarded as large-scale disasters, they may be examined as instances of near-miss disasters which offer examples of a particular configuration of pre-disaster communications. A collision may be regarded as an outcome of a two-party, non-zero-sum game, in which part of the process that produces the outcome undesired by both parties is a failure by one or both parties to interpret correctly the moves of the other. The parties to collisions may be unaware of each other's presence altogether, or they may miscalculate each other's behaviour.

The communication patterns associated with marine collisions, of which the above three cases may be taken as examples, display the most intensive communication flows *within* the two vessels concerned—intra-party, rather than inter-party communications. Inter-party communications are much less frequent, and while, in some cases, there may be radio communication between vessels, often those in command must rely upon visual, aural and radar observations of the other vessel to try to anticipate its moves. Communications within the vessel in this kind of accident may be ideal, but the difficulties of anticipating intent across a barrier of distance, using restricted communication codes, will still help to foster the development of a collision. Finally, it may be noted that such examples bring out the manner in which some accidents or disasters may arise as a result of the interaction of more than one set of purposive acts, since each party to a collision would have been perfectly safe had the other one not been there [17].

6.7. Discussion

We may now offer some general comments on the patterns of disclosure relating to disruptive events which are outlined in the above cases.

Routine or institutionalized communication patterns may be expected to grow up within any social group, as part of the normal interaction of everyday life. Also, such patterns become established between groups of individuals when there is an occasion for contact and interaction or an accepted need for the transmission of messages between the groups, whether these messages are concerned with tasks or with other matters [18]. Within work-groups, patterns of shared values, norms and perceptions tend to emerge, leading to the development within organizations of cultures and subcultures which the members of the groups hold in common [19].

The existence of such groupings also has implications for the process of decision-making, for the shared expectations associated with clusters of roles in an organizational group tend to encourage members of the group to bring certain assumptions to the task of decision-making within the organization, and to operate with similar views of rationality [20].

Patterns of shared culture and shared assumptions about decision-making are essential for the functioning of any organization, since they make for economies of communication. They ensure that certain types of information, which the groups concerned handle frequently and routinely, can be transmitted, received and understood without the need constantly to define all terms, or to justify basic assumptions before communicating the simplest message. But, by the same token, the internal communication systems of such groups will not readily handle messages based on a 'language' or a set of language assumptions differing from the one which is internally current among the groups; unless they are able, as Burns has suggested in his study of electronics companies, to make use of 'translations' when messages are being transmitted across boundaries [21].

Groups within organizations, then, develop their own cultures, and their own communications systems which are permeated by the assumptions of the culture within which they develop. Moreover, since the logic which lies behind the establishment and maintenance of many groups within organizations is the logic of the distribution of labour, and of the sharing out of tasks, the patterns of communication within an organization may be expected to bear a close relationship to the patterns of interaction established by the distribution of tasks within that organization [22]. Since most tasks tackled by organizations are large in scale, and require the co-operation and collaboration of several work-groups for their successful execution, routine patterns of communications grow up between groups.

In the light of these very general properties of the communication systems which develop within organizations, it can be seen that when failures, difficulties, errors or unforeseen hazards arise, they will be spotted more readily if they can be detected by the use of existing communication channels, and they will be dealt with more effectively if information about them can be passed to the appropriate executives through existing channels, as a matter of 'urgent routine'. But when difficulties or dangers arise which can only be fully appreciated or dealt with as a result of communications which must take place outside the normally established channels, the chances of an early alert and prompt evasive action became more remote.

It was suggested above that one basis for the formation of groups within organizations, and thus for the formation of communication patterns, is the distribution of tasks within the organization. But there are other bases upon which groups may be formed, so that other factors influence the development of concentrations of individuals who share to some degree a common culture within an organization, and who are accustomed to passing certain kinds of information amongst themselves in a rapid and routine manner. The informal structure within an organization provides for other bases of association than those formally prescribed by the authority structure of the organization. Some subcommunities may become established simply because their members are accustomed to being near to each other during the working day. Other groupings within organizations develop as a result of the formation of cliques, which in its turn may depend upon age criteria [23], upon the existence of strong kin relationships [24], or upon shared membership of religious or other voluntary organizations [25].

Since formal and informal groupings are not mutually exclusive, it is quite normal for individuals to be members of more than one such group, so that organizations are permeated by the interrelationships created by these numerous and overlapping groups. Extensive overlapping chains of relationships of this kind provide the means for the internal exchange of information on an informal basis which is known as 'the grapevine' [26], and it is also possible for 'the grapevine' to extend beyond the formal confines of a single organization and to link up with groups in neighbouring organizations, or in the community. However, the important point to be noted is that, while the existence of such overlapping sets of informal groups may augment and add flexibility to the formal structure of the organizations concerned, they are not infinitely fluid, so that they have their own routines, and follow their own habitual procedures. Thus, the structured nature of the communications, at informal as well as at formal levels, may act to facilitate the discovery of certain kinds of hazardous situation, but may hinder the recognition of situations which demand the flow of information in a non-habitual manner if they are to be discerned.

The specific point which arises here in relation to the seven case-studies set out above is that, whilst all disasters appear to arise as a result of the accumulation of unnoticed, misunderstood or miscommunicated events within the incubation period, a particular sub-class of disasters arises because of the barriers to certain kinds of communication erected by social habit and routine intercourse. Patterns of communication are structured by authority and the division of labour, by territorial considerations and by the formation of cliques, and where these institutionalized patterns create information-flow difficulties which result in the accumulation of unnoticed or misunderstood events, a pre-condition for disaster is established. Often, this condition is only ended by the occurrence of a widely noticed and widely recognized precipitating incident, such as an explosion or a collision. But, less frequently, it may be suspected, messages which for some reason are transmitted across the institutionalized barriers to communication may serve to check a developing disaster, as happened in both phases of the London smallpox outbreak.

It can be seen from an examination of Figure 6.5 that only in the first case discussed was the hazard noticed and dealt with within an area

ERRORS AND COMMUNICATION DIFFICULTIES

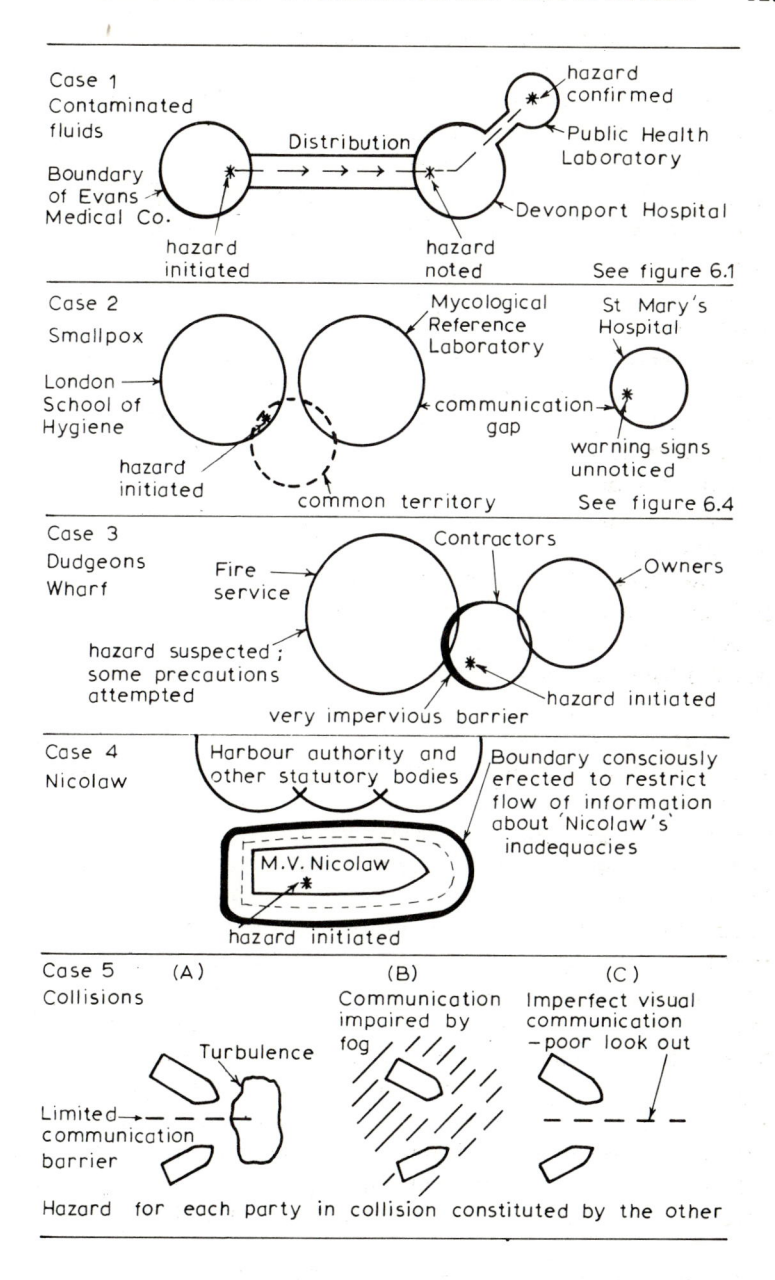

Fig. 6.5. Hazards and communication boundaries.

already established by means of routine communication patterns and communication boundaries. In the case of the smallpox outbreak, the Dudgeon's Wharf explosion and the loss of m.v. *Nicolaw*, communications barriers existed, separating those who were aware of the existence of the warning signs from those who could 'decode' such signs and act upon them. The last three collision instances also arose as a result of the difficulties of communicating in a precise manner across the barrier which a stretch of sea between two ships constitutes. It is not possible to identify unequivocally the point at which the hazard itself, or the warnings of the hazard, arose in the diagrams referring to these three cases, because in the particular instance of accidents arising as a result of the acts of two parties, the hazard for each party is constituted by the other party.

It may be concluded, then, that the nature of the communication patterns and the barriers to communication which prevail during an incubation period are likely to be of particular interest to those concerned to study the origins of disaster. The above analysis also serves to draw attention to a certain over-simplification which was displayed at the start of this chapter. Figure 6.5 carries with it the implication that there is a degree of social homogeneity surrounding the development of disasters, in that, before the disaster, everyone is equally misinformed or ignorant, and afterwards, everyone is equally well informed. For many, and perhaps most, instances of disaster, however, this simplifying assumption does not strictly hold.

The sequence of phases presented in Table 5.1 relates readily to the situation of an individual who collects clues relating to a new and, at first, unsuspected situation before it develops into a surprise or a shock for him; and, by extension, it can be related also to any groups which are sufficiently close-knit or efficient in their internal communications to be regarded as being like a single individual. In such cases, while disclosure may be inhibited because certain events are unnoticed or misunderstood, the implication is that understanding them or noticing them produces disclosure. The later discussion in this chapter should have made it clear that these implications will often be wrong. The student of disasters must seek the aid of the sociologist, to help him to discern those points in the social structure at which elements of the incubation pattern are potentially available, and those at which barriers to perfect information flows are established as a result of the operation of institutionalized communication systems.

And beyond this it becomes necessary, also, to start to take account of those other perennial concerns of the sociologist, the charting of the distribution of power, of the control of resources and of social reputation. The distribution of all of these elements, together with the actions of those who seek to change their distribution, may lead to intentional or unintentional influences upon the nature of the information flows which characterize the incubation period, prior to the disclosure of an unsuspected and dangerous situation.

6.8. *Summary*

Using the evidence from a variety of cases of accidents and disasters, it is possible to fill out the stage classification for the development of disasters already presented, and to suggest that the development of an incubation period prior to a disaster is facilitated by the existence of events which are unnoticed or misunderstood because of erroneous assumptions; because of the difficulties of handling complex sets of information; because of a cultural lag which makes violations of regulations unnoteworthy; and because of the reluctance of those individuals who discern the events to place an unfavourable interpretation upon them.

The development of the incubation period is further facilitated by difficulties of communication between the various parties who may have information which, if redistributed, could end the incubation period. The seven cases discussed, selected from the total of 84 cases listed in the Appendix, provide illustrations of the manner in which communications may be inhibited by institutionalized communication patterns which relate to physical or organizational barriers, or to the existence of various forms of social grouping. The latter section of the chapter draws our attention particularly to the manner in which it is necessary to pay attention to the influence of power, prestige and status upon the flow of information and upon the development of disasters.

7. Order and disruption

The analysis of 84 disaster and accident reports set out in the last three chapters, and in the Appendix, has made it possible to build upon the work of earlier writers in order to present a theoretical view of the events preceding disasters, in which there is a stress upon the importance of information, communication and disclosure patterns within the pre-disaster period. These patterns are closely associated with the production of the unexpected quality of surprise which is characteristic of accidents of all kinds. We now embark upon a more abstract discussion of some of the issues which our attempts to analyse the data have thrown up. In order to extend our understanding of the events and preconditions associated with disasters and accidents over the eleven-year period with which we are concerned, in this chapter and the two following ones, we shall introduce a number of general ideas which can be linked to our earlier discussions.

It is not possible to consider disasters for very long without running into questions about the nature of order and disorder, and some general properties of order and of the kind of disruption brought about by disaster will be reviewed in the first sections of this chapter. Then, since modern man usually sets out on his search for order with some kind of rational model of foresight and action in mind, we shall also look at some of the characteristics of rationality before going on to consider the nature of the unintended consequences which occur when rationality fails and order is disrupted.

7.1. *Purposiveness and order*

We can only be aware of the occurence of a disaster, or indeed, of any form of accident, if we have in our minds some idea of an alternative and more orderly pattern of events which would have existed had the accident not occurred. All accidents and disasters are measured in terms of the disruption of an order which was intended or at least anticipated in the future, and for this reason it is useful to look a little more closely at what is meant by terms such as 'order'.

Whether we like it or not, action can only take place in the present, but whenever men act voluntarily, or purposefully, their actions are oriented to some extent towards the future, to the time when they expect to see some outcome or achieve some goals as a result of their actions. Whether or not what they intend is attainable, their action generally presupposes that there is some possibility of achieving their intended purpose in the future.

There does not necessarily have to be an explicit goal which is clearly and fully formulated before the action is taken, and indeed, all life forms, including those with no evident form of consciousness, display a purposive element in their behaviour and organize their present behaviour as though they were attempting to realize some end in the future. Many writers have pointed to this purposiveness as a crucial characteristic which distinguishes living from non-living matter, the purposiveness of living matter being expressed in its tendency to develop and extend orderly patterns through time, in contrast to non-living matter which, if left to itself, tends to disorder and randomness [1].

A term which is often used to describe this tendency towards order in living matter is 'negentropy'. This derives from the word 'entropy', which originally had a very limited and specific meaning in thermodynamics, but which has been used more generally to refer to the tendency of non-living matter to move towards a random, disorderly state as a result of chance processes. In this more general usage, living organisms may be seen to have an opposite tendency, developing orderliness and acquiring or 'feeding upon' negative entropy or negentropy. This use of the term is due to Schrödinger, the theoretical physicist, who referred to the order-seeking tendencies of living matter in the following way. 'The device,' he comments 'by which an organism maintains itself stationary at a fairly high level of orderliness really consists in continually sucking orderliness from its environment' [2].

In spite of the fact that it is necessary to exercise some caution in transferring these terms 'entropy' and 'negentropy' from the original thermodynamic context in which they were formulated, they have gained increasing currency in the fields of biology, information theory [3] and social science [4], and since it is convenient to use them to refer to tendencies towards increasing disorder and increasing order respectively, they will be used in this manner in the following chapters.

All life, then, has a purposive quality, but there is a distinction to be made, as has already been suggested, between purposive behaviour unconsciously directed towards the completion of a process, even where this involves persistence towards the goal in spite of deflections, and purposive behaviour where the attainment of a goal is *seen* as a purpose, and a solution is consciously considered [5]. There is a difference, that is to say, between behaving purposively, and knowing the goal towards which the purposive behaviour is directed. All life forms behave purposively, but knowledge of a goal can only be associated with consciousness, and this therefore seems to limit such behaviour, at the moment, to those with human consciousness. Human behaviour thus displays both forms of purposiveness, for in addition to pursuing conscious goals, human beings appear to be born with a sense of hope, or of invulnerability [6], with a kind of blind feeling that the advancing spread of entropy will continue, somehow, to be overcome.

But if goals are to be attained in the future, it is not enough merely to have an intention of reaching them. People never act in a vacuum: they must achieve their goals in the environment in which they find themselves, and their success or failure depends to a great extent upon the degree to which their environment is congenial to their goals. If their surroundings are arranged in a convenient manner, and if their fellow humans are helpful, they may be successful in reaching their goals, but if the 'munificence of the environment' [7] takes a downward turn, the eventual attainment of the desired ends will depend upon the resources they have available to cope with the difficulties, and upon the degree of knowledge available to them about the nature and location of possible hindrances.

We do not need to have a perfect or complete knowledge of the environment in order to be able to act in it, and to move towards goals which we desire. All that we need is knowledge which is accurate enough for our purposes, although we can never say with absolute certainty that we do know enough until after the event. Instead of perfect knowledge, the human organism operates with a variety of maps and models which organize available knowledge about the outside environment, and direct the collection of new information. These guides for conduct range from physiological 'models' built into the body to regulate breathing according to the oxygen content of the air, to mythical models which organize information about the natural environment in order to help man to preserve an ecological balance [8]. The problem of determining just *what* level of accuracy of knowledge is required for goal attainment, or indeed for survival, is one which can never be finally solved however—for all maps, models or other bases for organizing information inputs have assumptions built into them about the relevant portions of the environment which must be attended to, and about the portions which may be ignored with safety. Major shifts in the environment may render existing maps useless, whether they are physiological, mythical or cognitive, and it is the resulting discrepancies between the way in which the world is believed to be, and the way that it really is, which contain the seeds of disaster.

7.2. *Rationality in the pursuit of goals*

In Western civilization, the models which are built to guide actions in the world are likely to be 'rational' models [9]. Rationality is an elusive concept [10] but, in general terms, we might expect rational models in Western society to be constructed with the intention that they should be internally consistent, and specified with a precision which enables relationships within the model to be calculable, in principle. Similar requirements should apply to the relationships between the model and the environment to which it relates, so that when the model is used as a guide to purposive behaviour, it enables consistent, predictable outcomes to be attained and eliminates magic as a means of securing ends. This kind of rationality is highly prized in our civilization [11], but it represents an ideal of orderly, predictable thought and action which is not always attainable; later on, a much more restricted view of rationality or intended rationality will be referred to [12].

Western man often assumes, particularly when he is organizing action with others, that he will be able to arrange and direct his actions by means of such elaborate, consistent and systematically rational maps of the environment, which he constructs (in his head or on paper) to help to make the outcome of his actions more predictable. He has clearly achieved some success in the construction of such maps, because his civilization depends upon them. Without satisfactory plans, models or maps, using these terms in a wide sense, it would not be possible to construct or maintain the complex systems of defence, sewage, transport, energy generation and distribution on which his society depends.

Merely being able to organize resources and knowledge in a rational manner which is directed to the attainment of a goal does not, as a matter of course, guarantee that the goal itself is desirable or worthwhile. The appearance of rationality may be deceptive, as a number of writers have been concerned to point out [13]. The sociologist, Mannheim, makes this point by distinguishing between what he calls 'functional' and 'substantial' rationality. He describes functional rationality as a series of actions 'organized in such a way that they lead to a previously defined goal, every element in this series of actions receiving a functional position and role' [14], and he goes on to observe that 'the more industrialized a society is, and the more advanced its division of labour and organization, the greater will be the number of spheres of human activity which will be functionally rational, and hence also calculable in advance' [15].

But he is careful to distinguish this kind of calculable activity from 'substantial rationality', which arises from an act of thought revealing 'intelligent insight into the interrelations of events in a given situation' [16]; and he points out forcibly that, merely because industrial society displays a large amount of functional rationality in all of its aspects, it is not necessarily a society with a great amount of substantial rationality. Indeed, since large organizations and extensive technical systems subdivide thought about problems and fragment the responsibility for them, many individuals within such systems are likely to be deprived of the need to develop their own thoughts, insights and responsibilities. Only a few individuals in key positions in society are expected to think intelligently and independently [17]. At times, as we have seen, the inability or the unwillingness of individuals concerned with routine jobs to question the logic of the system to which they contribute may lead to disaster.

It is clear, therefore, that rational behaviour has its limitations, particularly if we are misled into accepting functional rationality as a substitute for substantial rationality: the existence of a large organization, with an impressive organization chart, elaborate corporate plans and an extensive staff of specialists offers no guarantee of substantial rationality. And because it is important to become aware of such discrepancies, Mannheim hoped that his analyses would increase the awareness of them, and encourage concerned individuals to devote some of their efforts to promoting a wider spread of substantial rationality in contemporary society. One of the ways in which this can be done is by maintaining a critical stance towards the models which large, intendedly rational organizations use to guide their actions, and the present book may help to offer a basis for the formulation of criticisms which may help to transform functional rationality into substantial rationality. Clearly, the more substantially rational a society is, the less it is likely to be afflicted by disaster.

If men knew everything, there would be no discrepancy between functional and substantial rationality, between intention and outcome, for everyone would always have the knowledge to realize their intentions. Their 'radius of foresight' as Mannheim called it [18] would always exceed their 'radius of action'. It is obvious that in such a world, many specialists would find themselves redundant, not least the sociologist, for it has been suggested that the study of unintended consequences is one of the central concerns of sociology, and the sociologist would thus find himself deprived of his subject-matter. However, in reality, the 'radius of foresight' is much shorter than the 'radius of action', accidents and disasters occur frequently, and the study of unintended consequences must clearly retain a central place in any attempt to understand the nature and origins of these incidents.

The lack of knowledge which leads to unintended consequences may relate to many aspects of the environment: material, technical, social and cultural; and a variety of theories of notably differing kinds, ranging from Marxism to functional analysis, have included the idea that man's deliberate attempts to shape the future are limited, constrained and frustrated, not only by nature—and by nature modified by technology—but also by the existing structures of society and by existing cultural patterns [19]. It is not only technical ignorance which hems us in, for our inability to foresee even the most important outcomes of changes which we may make to the social structure severely limits our ability to intervene successfully in social as well as in technical affairs, and there seems to be little prospect that we shall be able to improve our capacity for foresight in the social field significantly in the near future [20].

Unintended or unanticipated consequences follow from individual or collective acts, then, because of man's lack of full knowledge of the implications of his acts, although it is by no means always the case that consequences will be undesirable merely because they are unanticipated [21]. Some early analyses of the reasons for the limited nature of the knowledge available for the guidance of action tend to stress limitations from the point of view of the individual, although the analysis of disasters presented earlier, as well as a consideration of the 'functional/substantial rationality' distinction, would point also to the importance of wider social factors. Evidently any satisfactory analysis should include both the individual and the group factors which limit knowledge.

Why should foreknowledge be limited? In some cases, of course, there is a complete ignorance of events in the incubation period, so that the outcome cannot be foreseen at all, but the analysis developed earlier in this book was predicated upon the assumption that very few disasters arise from such complete ignorance. Where no foreknowledge is available at all, it is not possible to carry out any kind of scientific inquiry and the adoption of a fatalistic attitude is the only available option.

Knowledge may be thought of as being limited in two ways, although there is no hard and fast separation between the two. Before we embark upon a course of action, when we try to work out its implications in advance, we may be limited in the extent to which we can foresee the full range of consequences which will develop; and then, once we have embarked upon a course of action, for a variety of reasons we may fail to be aware of the untoward turn which events are taking, even though clues and signals about this may be available.

The first case arises because of the inadequacy of our ability to predict accurately, either because our models and theories about the future are inadequate, or because the necessary information which must be inserted in them is not available. Thus, we may be unable to predict outcomes accurately because we make errors in appraising the present situation, or because the information which we can gather is only available in a limited form, in probability terms, say, which may not be very helpful to us in dealing with a specific case. Or, very commonly, we may be initiating action in a complex situation, where the variety of possible interactions which may affect the outcome which we intend may be great, so that our forecasting ability is limited. We may not be able to spread our attention sufficiently widely to take account of all the developments which we initiate, for while we are concentrating upon the immediate ends in view, some of our actions may affect adversely the ultimate goal which we have in mind [22]. There is a problem, that is to say, of 'keeping an eye on the ball'. In the social realm, particular forms of this difficulty arise as a result of phenomena like 'self-fulfilling prophecies', when events occur because they have previously been predicted [23], or as a result of the opposite effect, where the overt prediction of an event inhibits its occurrence.

Another important factor which limits our ability to predict the outcomes of our actions relates to what Selznick has called 'commitment'. In a study which charted the factors which prevented the American New Deal agency, the Tennessee Valley Authority, from fully reaching its intended goals, he drew attention to the way in which the need to commit oneself to a course of action could operate to limit one's options at a future time [24]. Adherence to any set of practices or rules, or the employment of any group of personnel, limits the perception of available alternatives, and limits, too, the freedom of action to pursue such alternatives even if they are seen. This constraint may be due to commitment arising (i) from the particular patterns of order or discipline adopted within an institution, (ii) from the social character of the personnel engaged in any enterprise or institution, (iii) from habitual and accepted practices which become established within an institution or organization, (iv) from the nature and restrictions of the social and cultural environment within which action is taking place, and finally (v) from the centres of political interest which may be generated by or affected by an action. Selznick's study provides extensively documented instances of the way in which organizational intent can be diverted as a result of all five of these reasons, and we may expect to find similar processes influencing the intendedly rational actions of those individuals acting within organizations before large-scale accidents or disasters. The second general manner in which knowledge may be limited which was referred to above occurs after a course of action has been embarked upon, when the signs of impending failure are potentially, but not fully, visible. This kind of limitation of knowledge is central to the definition of the disaster incubation period which has been discussed in the last two chapters, and the relevant kinds of limitation upon knowledge are those which have been reviewed there: knowledge is limited because of perceptual difficulties arising from distractions, anxieties, rigidities of belief, cultural lags and institutional constraints; or because of a range of difficulties relating to the problems of transmitting information rapidly and accurately from individual to individual or from group to group [25].

But whether accidents and disasters occur as a result of ignorance in the early stages when action is planned, or as a result of the neglect, for a variety of reasons, of warning signs while a course of action develops, there are certain constraints upon the ability of individuals to gather and process information which are of general relevance to this discussion. These can best be described by outlining some of the points developed by Herbert Simon in his work on the nature of administrative rationality.

7.3. *Bounded rationality*

We may begin with a distinction which Simon draws between 'limited rationality' and 'global rationality' [26]. To the extent that it draws attention to some of the boundaries of rationality, this distinction is similar to the 'functional/substantial' classification which we have already discussed; but Simon is less concerned with wisdom and the appearance of wisdom, than with much more mundane matters relating to man's limited ability to search for, collect and process large amounts of information. Consequently, Simon's observations on these matters apply to the administrative capabilities of individuals within functionally rational organizations, just as much as they do to those in substantially rational organizations.

Simon relates his distinction between 'limited' and 'global' rationality to a principle which he calls the Principle of Bounded Rationality, in which he states:

> ... that the capacity of the human mind for formulating and solving complex problems is very small compared with the size of those problems whose solution is required for objectively rational behavior in the real world ... or even for a reasonable approximation to such rationality [27].

Thus, even if man manages to avoid a pure 'functional rationality', and is attempting to pursue reasonable and appropriate goals, even if he sets on one side his propensity to act from irrational urges, and his inveterate tendency to clothe the pursuit of inappropriate goals with the appearance of reasonableness, he does not escape the bounds of 'limited rationality'. We may take Simon's Principle to indicate that even when men are behaving 'substantially' rationally, rationality may fail to deliver the goods, because it is a mode of human thought and humans are finite.

The limits on rationality are essentially the limits of information-gathering in a world where information-gathering has costs; the limits of the number of possible alternative solutions which can reasonably be considered and evaluated, in a world where many of the most important problems have an infinite number of possible solutions; and the limits of a lack of knowledge of future events, where rational behaviour depends upon the formulation and successful execution of plans which promise attainments in the future.

Simon handles this issue [28] by first setting up a model of perfect rationality, and then showing how the kinds of constraint mentioned above limit its operation. He envisages the individual being confronted with a set of possible alternatives for action, although the individual may only perceive or consider a proportion of these. Each possible course of action may be thought of as bringing about a future state of affairs. Or, in probability terms, there may be a probability distribution of the likely future states of affairs to be attained if a particular alternative is chosen.

The perfectly rational person will choose the alternative leading to the best possible outcome for him. But a real decision-maker, because of the limits implied in the Principle of Bounded Rationality, will be forced, in any but the simplest of situations, to abandon the pursuit of a maximizing strategy, accepting a satisfactory outcome rather than the best possible one. Simon refers to this approach as 'satisficing' rather than maximizing [29], and we may think of it as a result which is achieved by evaluating outcomes as 'acceptable' or 'unacceptable', or as 'win, draw or lose' outcomes. As a result of having to count the costs of gaining more information against the costs of taking action based upon present knowledge; as a result of having to assess outcomes in vector terms, where the output of a decision may be measured in two, three or more incommensurable measures; and as a result of being unaware of possible alternatives of action lying outside those considered, the decision-maker is forced to settle for 'acceptable' or 'win' outcomes (if, indeed he is able to achieve these) rather than pursuing notions of *maximum* acceptability, or *best possible* win.

It is never completely clear [30] whether Simon's decision-makers are required to be conscious of their limited rationality or not, but in the present discussion it makes sense to assume that all decision-making is limited, whether the decision-maker is conscious of it or not. And, in any case, it would seem likely that any decision-maker who had read his Herbert Simon and was aware of his own limited powers would nevertheless be incapable of perceiving the exact boundaries of his own limited rationality.

An administrator can, of course, try to extend his range of effective action, and Simon discusses a number of strategies which he may adopt in order to do this. One strategy is to make exploratory trials in his early decisions, and try to learn from them for future decisions, so that he is not confined to the highly artificial case where he has to make one selection from a single array of options facing him. Or he may choose to base his present decisions on known past decisions, using the strategy of 'incrementalism' [31].

Where his past decisions have been satisfactory, and appear to offer a good basis for future decision-making, each successive round of decision-making increases the information about, and raises his level of certainty with regard to, the consequences of the known alternatives and their variations, and the range of alternatives he looks at is likely to remain small. If, on the other hand, it is difficult to find satisfactory solutions, the more persistent the decision-maker, the more he will seek to enlarge the set of considered alternatives. And, as Simon also points out, there may possibly be some higher level of costs, values and rational considerations which offer a guide as to whether to be more or less persistent in these searches, although such rationalities are themselves subject to all of the above strictures. This hierarchical notion of successive levels of rationalities will be developed in the next chapter, when we come to consider the relationship between processes of organizational decision-making and the emergence of hazards [32]. For the moment we may merely note that, although Simon's discussions are couched primarily in terms of the limited information-processing and decision-making abilities of the single individual, the individual concerned is an administrator, functioning within an organizational setting, so that a close interrelationship between individual and collective decision-making processes is implicit throughout Simon's analysis.

The strategies outlined above indicate ways in which the individual decision-maker can attempt to behave rationally, by searching for courses of action leading to 'satisfactory' outcomes. There is no way in which the individual decision-maker can guarantee a 'best possible' outcome, unless he is acting within the limits of a small closed system about which he can gain perfect knowledge. Even then, since all closed systems are abstractions, any *real* decisions can only deal with open systems.

Any problem of information-collection or information-processing is not just a subjective problem for the individual administrator: it is also a collective problem for the institutional grouping within which it occurs [33]. An organization has a much greater capacity for gathering and processing information relevant to action than any single individual; but, even so, this capacity is still finite and is still too small to cope adequately with the solution of many of the problems that an organization is likely to be faced with. Simon's Principle of Bounded Rationality points to the limits of effectiveness placed upon both individual *and* collective decision-making which ensure that there will always be an area of potential ignorance which may serve, at times, to turn man's best intentions into disasters [34].

The idea of bounded rationality is important, therefore, in exploring the nature of unintended consequences, and in searching for the origins of disasters, and we shall return to it in subsequent chapters. It is not sufficient, however, merely to note the principle in abstract, and we shall pay particular attention to the manner in which bounded rationality operates within hierarchical organizations, and look at the kinds of event associated with the failure of bounded rationality.

7.4. *Summary*

Disasters and accidents are measured by men against their intentions that order should develop and extend into the future. Western man in particular has attempted to pursue his intentions by the development of elaborate and extensive rational schemes and plans for action which he believes will help him to subdue and control the world. In spite of the fact that such rational constructions ignore or belittle the intuitive and the emotional aspects of man's life, they have a considerable degree of effectiveness, and our civilization is dependent to a great extent upon large-scale organizations which embody such ideas about action.

Rational plans for action, however, do not provide perfect guides: they may be 'functionally' rational plans, with all of the outward trappings of intelligence directed towards the pursuit of a goal which it would have been wiser to avoid; they may be based upon inadequate foreknowledge of the forces likely to affect the proposed actions, or of the interrelationships between these forces; and the ability of the individuals or groups concerned to collect and process the information required for the accurate guidance of action may be limited by the finite nature of all human decision-makers. There is thus, ultimately, no way in which unintended consequences can be avoided with certainty, so that the potential for disaster is always with us.

The discussion presented in this chapter has been based mostly upon a consideration of the individual decision-maker within an administrative setting; but, because of the interrelationship between individual and collective ignorance, more attention needs to be paid to the effects which an institutional setting has upon the decisions taken within it; for this reason, in Chapter 9, the way in which the ideas discussed above apply within a hierarchical organizational setting will be considered.

8. Information, surprise and disaster

An individual acting with a sense of purpose, with an intention of achieving some end, initiates a series of actions according to his understanding of the operations necessary to reach his goal. If he is to be successful, his intentions will be paralleled by a sequence of events in the world which achieve what he had in mind, within limits acceptable to him. But if he misunderstands the world, at some point the events which he initiates will diverge from the path which he intended for them. This divergence may continue for a few seconds or for many years; but, at some point, his misunderstanding of the world in which he is trying to operate will become evident when the outcome, the sequence of actual events, fails to coincide with the intended sequence. At this point, the incompatibility between his plan and the environment within which he is trying to realize this plan becomes evident, as an accident, as a disaster or as some other form of unexpected and unintended outcome.

Men's plans are checked whenever they come up against the limits of their bounded rationality; and if we are to understand the nature of accidents and disasters, and the manner in which they come about, it is clearly essential for us to understand what happens when a limited rationality becomes over-extended. In this chapter we shall pursue this question at some length, raising a number of fairly abstract issues relating to the nature of information and knowledge; and, in the following chapter, this new-found understanding will be set into the organizational and institutional contexts which have already been found to be of central importance in the development of disasters.

8.1. *The nature of information*

To gain an understanding of the abrupt changes in information-flows which the surprise of an accident or disaster brings about, it is useful to review some general properties of information, paying particular attention to the manner in which information is dealt with in the discipline known as 'communication theory' or 'information theory' [1]. In this body of theory, stress is placed upon the need to obtain measurements of information and of the information-content of messages in various situations. The accepted means of measuring the information-content of a particular message is by assessing the reduction in uncertainty which the receipt of a message produces for the recipient, and mathematical techniques for specifying this amount precisely have been developed [2].

This idea of information-transmission and uncertainty-reduction, however, is not a simple matter, for there are several complexities associated with its application. The means of measuring information and information-content were devised essentially to provide tools for use in communication engineering, and they relate directly to the problem of assessing the message-carrying capacity of a communication channel, and of measuring the efficiency with which messages are sent along such a channel, from transmitter to receiver [3]. One very relevant property of such an arrangement is that the message is normally assumed to be transmitted in terms of a set of categories known to both sender and recipient. Thus it may be accepted, for example, that a given message will be transmitted along a telegraph wire, in Morse code, in English, so that before the message is sent the individual waiting for the message knows the categories available for transmission. What he is uncertain about is which of these categories will be transmitted, and the information-content of the message is then measured as the extent to which it reduces this uncertainty for him.

The notion of information which is central to information theory, then, is a very precise and restricted one, which is measured in terms of the change it produces in the level of uncertainty experienced by the recipient of the information. Although this point is normally irrelevant for communication engineering applications of information theory, it may also be noted that such an approach to the measurement of information implies that uncertainty-reduction can only be assessed in relation to the existing amounts of information already held by the receiver. For measurements of, say, the information-carrying capacity of a long-distance telephone line, this point is of little moment; but if we are interested in transferring the ideas derived from information theory to a slightly wider context, questions about the prior state of knowledge of the recipient of a message become much more central; for example, for a motorist, each gradation on a speedometer with a range from 0 to 100 miles/h will not have the same potential information value for him, since he will not normally be in a completely uninformed state about his speed before looking at the instrument [4]. The motorist may think, say, that he is probably travelling at some speed between 30 and 50 miles/h, and that the probabilities that he is stationary, or that he is travelling at 100 miles/h, are both zero.

There will therefore be a distribution of probabilities which characterizes his knowledge of his likely speed before he looks at his speedometer [5]. This notional distribution will cover all of the likely or unsurprising outcomes, and new information will serve to resolve matters within that distribution. The universe of possibilities which he envisages, then, adds up to the certainty of his present situation.

Only information which will change the shape of this distribution can be said to affect our motorist's present uncertainty, for information does not resolve uncertainty unless it is appropriate information and is given to those who are uncertain. Thus, the information content of a message transmitted to this motorist which we could represent as 'You are not travelling in the speed range 80–90 miles/h' would be identical in communication-engineering terms to the content of the message, 'You are not travelling in the speed range 30–40 miles/h.' However, to anyone interested in the reduction in uncertainty experienced by this particular motorist, it is clear that the second message will reduce his uncertainty more, and will effect a greater change upon his understanding of the likely speed at which he is travelling than will the first [6]. Another way of expressing this is to say that he would be less surprised by the first message than he would by the second: it is no news to him that he is not exceeding 80 miles/h but his higher expectations that he was travelling within the slower speed range of 30–40 miles/h have now been dispelled.

This example raises the question—one very germane to discussions of accidents and disasters—of the relationship between information and surprise. To the extent that any resolution of uncertainty cannot be accurately predicted, it has been suggested that surprise is the essential feature of all new information [7]. But whilst this is true to a degree, we want to retain the possibility of distinguishing between two kinds of surprise, and thus of distinguishing between two kinds of information-interpretation—by separating out on the one hand the low level of surprise experienced by our motorist who thought that he was probably travelling at between 30 and 50 miles/h and now discovers that he is travelling at some speed between 40 and 50 miles/h, and on the other hand the kind of surprise that he would experience if he suddenly discovered that he was travelling at 180 miles/h.

Let us try to understand how such a distinction, which is central to an understanding of surprise and disasters, relates to the 'information theory' or 'communication theory' model we have already referred to. In such theory a communication channel, with its transmitter, receiver and given set of categories for message transmission, constitutes a closed system; and within this system the sum of the probabilities that information transmitted in message form will fall into the available sets of categories is unity; while the probability that information transmitted will fall outside these recognized categories is zero, from the point of view of the receiver. Thus, information which appears to fall outside these categories is readily regarded as error, arising from slips or distortions. This closed-system model is an excellent one for the discussion of problems of data-transmission, of message-redundancy and of the reduction of 'noise' which interferes with efficient transmission, and a number of writers would like to limit discussions of information to such a model [8].

In spite of such attempts to restrict its usage, however, the term 'information' is commonly used in a much wider sense, which blurs many of the precise technical definitions available [9]. In many cases, this vagueness is merely the result of the careless or casual use of the term, but there is also a strong and persistent factor which encourages a departure from the technical definition of information, and which is central to our theme. This is the habitual tendency, which human beings have, to regard the world as though it were full of messages for them, and to behave as though they were on the receiving end of a large number of communication channels originating both in the material world and in the social world.

A leading information-theorist, Cherry, has tried to resolve this problem by making a distinction between channels of communication and channels of observation. He points out that questions of extracting information from Nature and using this information to modify our theories about the world lie outside communication theory, so that an observer looking down a microscope or reading instruments is not to be equated with a listener on a telephone receiving spoken messages. 'Mother Nature does not communicate to us with signs or language' he says. 'A *communication channel* should be distinguished from a *channel of observation*' [10]. This is a most useful distinction, and we may go on from it to ask what the properties of a channel of observation might be, and to try to clarify the relationship between such channels and channels of communication.

We may first of all note that channels of communication in the narrow sense may become channels of observation if we step outside the closed communication systems already described. Suppose we wish to communicate with someone else, and our concern goes beyond merely wanting to ensure that he can hear and translate our words, to a desire to encapsulate within our message some information which he needs for action; then, as senders, we must face the problem of placing the information that we wish to send into appropriate classes in order to achieve our end (see Figure 8.1). In the same way, the scientific observer has to sort his observations into appropriate and meaningful classes, so that he can understand his 'messages from Mother Nature'. An adequate system of class assignment must therefore ensure that each item that we want to communicate about belongs to at least one of the categories in which messages can be sent. Unless this requirement is met, we may find ourselves debating whether an item is located at point A or point B, when in fact it is in neither place; or about whether we should keep our stocks constant or increase them, when they should be decreased [11].

When our classes exhaust the possibilities about which we wish to communicate, we have a stable situation to which we can begin to apply the considerations of information theory; but for observers of the world, and for those who wish to communicate in an innovative way to others

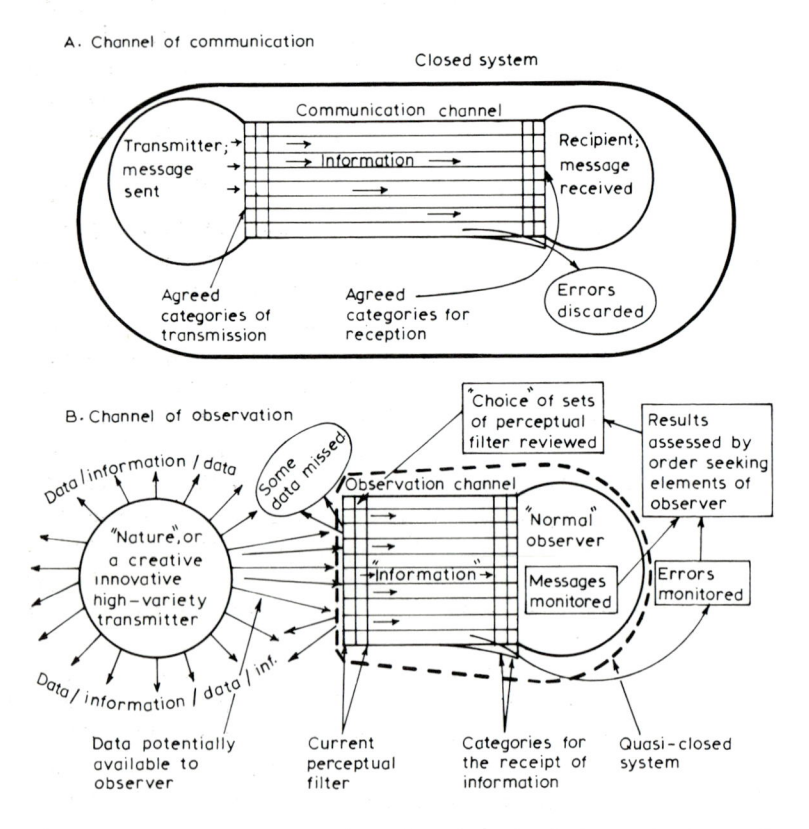

Fig. 8.1. Schematic representation of a channel of communication and a channel of observation.

about the world, the search for an appropriate set of classes in which communication can be carried out is a never-ending one; and there is a recurrent need to shift from one set of classes in current use to another set which should be more satisfactory in the future.

We can, perhaps, illustrate these general considerations by reviewing an example which seems to be particularly liked by information-theorists. When the celebrated American folk-hero Paul Revere and his friend agreed upon a light-signalling system to warn them of the direction of approach of the British enemy—'One, if by land, and two, if by sea;'—they did not *know* which of these routes the British would take to Lexington and Concord. They assumed, however, that they had an appropriate system of class assignment which reflected what they wanted to know about the behaviour of the British under three headings. No lights in the belfry of the church tower meant that the British had not yet begun to approach, while one or two lanterns meant that they were approaching, and indicated the route. When Revere saw a light across the river, his previous uncertainty was reduced, and as we noted above, we may regard this resolution of uncertainty as a low level of surprise. This case is discussed by information-theorists in introductory texts because it embodies the elements of a communication channel, with the sum of the probabilities of the three classes available for transmission being effectively unity [12]. That is to say, it was considered to be certain that the enemy would be in one of the three states which could be signalled.

When this assumption no longer holds good, however, the channel can no longer be regarded as a simple communication channel, and we need to extend our model to take wider account of the activities of the transmitter and the receiver. If Paul Revere's British opponents had started to attack in a fleet of hot-air balloons, the signaller in the church tower would have realized that the system for assigning events to communicable classes which Paul Revere had set up for him was now inadequate, and he would suddenly have become impotent as a transmitter of surprising but uncertainty-reducing information. He would have received, along the 'channel of observation' which linked him with the enemy, information which was incompatible with his previously accepted set of assumptions, information which in some senses increased rather than decreased his uncertainty, and he would be unable to pass this information on with the means of communication at his disposal (see Figure 8.2). He would be faced with the problem of coping with an event which had occurred, but which previously had been tacitly assumed to have a probability of occurrence of zero. We may thus regard this kind of event, which was not assigned a place in the relevant system of classes, as a

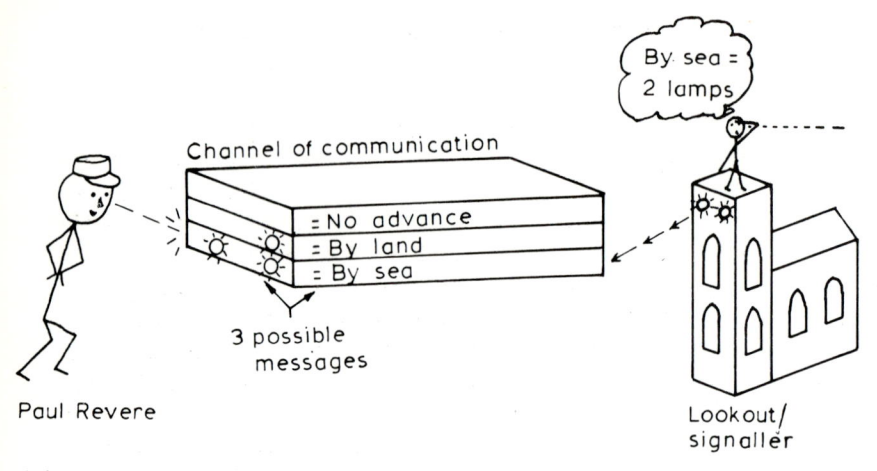

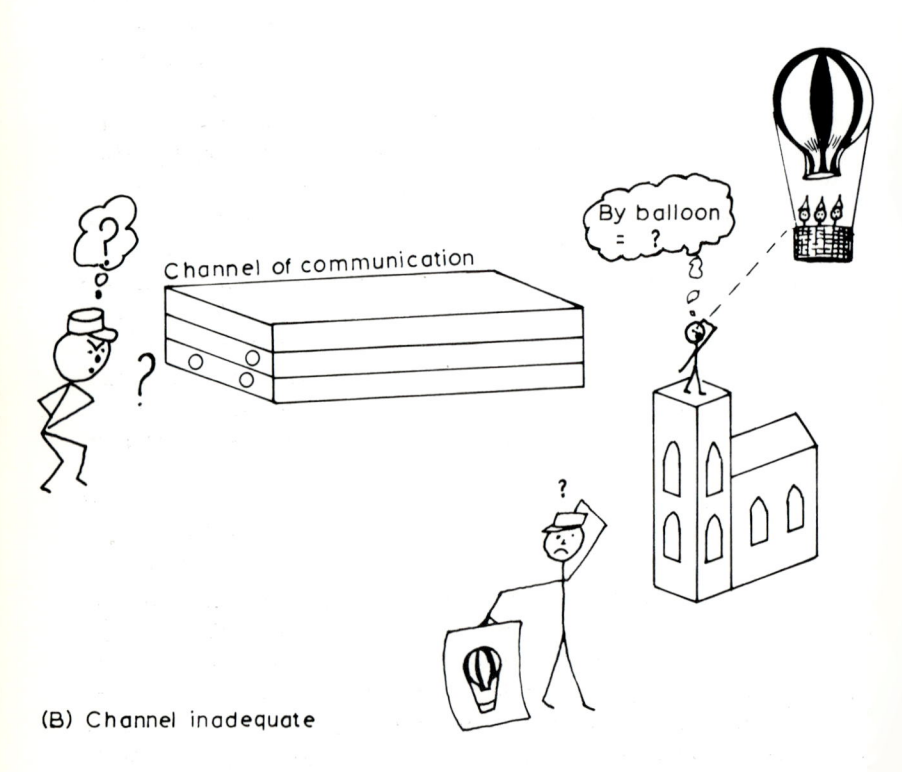

Fig. 8.2. Paul Revere: adequate and inadequate channels of communication.

source of a higher order of surprise than we could assign to events which could be dealt with within the existing sets of classes for perception and communication.

All accidents are unexpected, or surprising, but some are unexpected only in the sense that no-one knows exactly *when* the incident will occur, or *where* it is to occur. For such incidents, there is a readily available set of categories into which the message of the event's occurrence can be coded, and the news in this message, of the precise time and place of the accident, serves to reduce uncertainty. But where this is not the case, where the event is unexpected at a higher level, so that the news of its occurrence serves to increase uncertainty, it cannot be accommodated to existing understandings of the world without provoking considerable revision of these accepted modes of thought. This kind of surprise is occurring, therefore, if the receiver of information feels it necessary to revise the sets of categories within which he is prepared to receive messages. It is clear that we can only understand this kind of discontinuous shift on the part of the receiver by extending our attention beyond the limited number of components which make up a closed-system communication channel model. In particular, it is essential to include some assessment or consideration of the order-seeking propensities of active, inquiring observers as they monitor the processes by which they receive information along 'channels of observation'.

The nature of information, then, is evidently rather more complex than the definition of it in information-theory terms would suggest [13]. Information about events of all kinds is potentially available throughout the universe, but this information is in itself of no use unless it conveys a message to an order-seeking, negentropic entity. One of the properties of all such entities is that they constantly seek to absorb more negentropy, by decoding and taking account of more and more information. Random machines, on the other hand, cannot be 'informed' because information can only be absorbed by entities which are orderly enough to have already 'fixed' other information in some way [14].

This discussion implies that there is not a single level at which all information exists, so that we must think, instead, of a hierarchy of levels of information. If, as many nineteenth century thinkers appeared to assume [15], all information could be contained at some single level, any events which could not satisfactorily be accounted for would be indicative of either a failure to gather sufficient information, of an inadequacy of explanatory ability, or of both. This reductionist view of the world promised a 'philosophy of tidiness' with everything labelled and in its place, but the world cannot be reduced to such a single informational level, and the promise is illusory [16]. More adequate theories of life must take account of the possibility that information is ordered at many levels, and that there may well be an intrinsic absence of information at any given level [17]. The hierarchical manner in which information about the world must be arranged is important in gaining an understanding of the manner in which disasters and accidents are able to create surprise.

8.2. *Informational discontinuities*

When there is a shift from one level of information accumulation to another, the tacit and habitual assumption that all information must reduce uncertainty within one given set of categories breaks down, and a discontinuity appears in the uncertainty-reduction process. When a piece of information is acquired which cannot be ignored, which cannot be disposed of as error, and which cannot be accommodated within an accepted set of categories, the approximation of the channel of observation in question to a channel of communication no longer holds; and it becomes necessary, if faithful observation is to be continued, to establish a new set of categories within which both old and new information can be accommodated. After this point of discontinuity, when a new framework has been established, the uncertainty reduction process may once again be resumed.

The nature of this discontinuity in the uncertainty-reducing process may be more readily understood if a simplified model is discussed, for this will make it clear that two forms of discontinuity are possible. Let us imagine that an individual is gathering information relating to a particular topic, under the impression that all relevant information can be classified into, say, ten categories. As he collects information, his uncertainty about the messages which he expects to receive in those categories will decrease, by an amount which can be measured by a Shannonian or by some other suitable measure of information. But if, midway through this process, in the course of monitoring the way in which he is acquiring information, he realizes that there are actually thirty relevant categories, then his uncertainty will increase sharply before beginning to fall again as he collects further pieces of information. This process is represented diagrammatically in Figure 8.4, the simple situation where uncertainty decreases continuously with the acquisition of information being shown in Figure 8.3.

But it is also possible for our individual, beginning with ten categories, to realize subsequently that there are effectively, say, only five categories, and this second form of discontinuity is illustrated in Figure 8.5. In the first case the shift in assumptions made by the receiver meant that he

INFORMATION, SURPRISE AND DISASTER 147

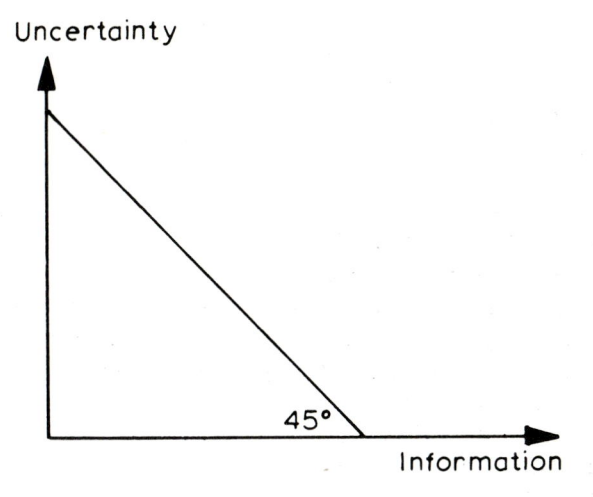

Fig. 8.3. Uncertainty decreases as information increases in a channel of communication.

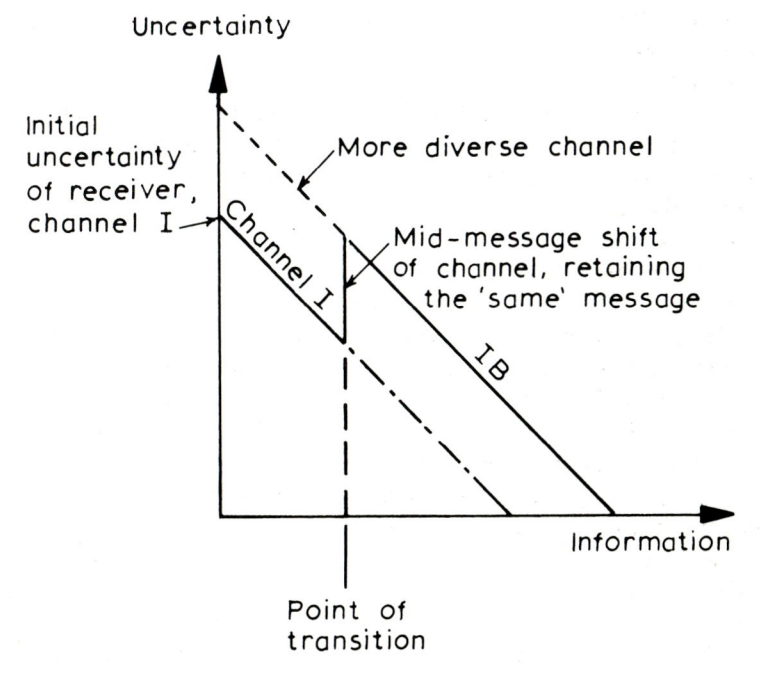

Fig. 8.4. Discontinuity in uncertainty-reduction as a result of a transition to a more diverse channel.

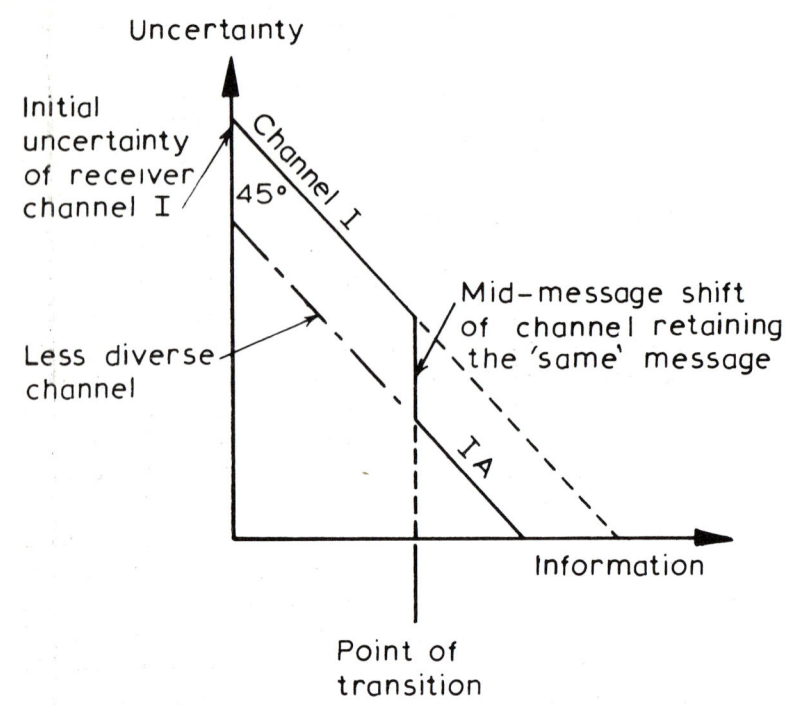

Fig. 8.5. Discontinuity in uncertainty-reduction as a result of a transition to a less diverse channel.

ceased to receive messages on the ten-category channel, and started to receive them on a thirty-category one; while in the second case, his shift in assumptions had effects similar to those which would be expected if he began receiving messages on a much simpler channel, with only five categories.

Of the first of these discontinuities, we might say that the receiver suddenly realized that the matter was not so simple as he had at first assumed; while in the second case he reaches a point of enlightenment at which he realizes that matters are really rather more simple than he had taken them to be.

Considering such simple illustrations, it is evident that in order to cope with discontinuous shifts of this kind, further departures from the closed 'communication theory' model must be made. Whether the receiver wants to jump to a more complex or to a simpler 'channel', it is clear that if the messages already received are not to be repeated, such 'mid-message shifts' must allow all of the information already received to be recoded so that it can be reinterpreted in the new categories; and for this to be possible, particularly when the shift is in the direction of greater complexity, it is evidently necessary that the receiver should have acquired and absorbed more information about each of the message units than was originally necessary to classify them, and this second stream of information must have been monitored in a critical manner. It can only be as a result of this second-level monitoring process that a shift from one channel to another can come about, for there is no element within the communication-theory model of communication channels which could initiate and carry out such a shift. See Figure 8.1.B. The realization of a need to change channels must be attributed, in short, to the order-seeking propensities of an active and intelligent receiver or observer who, in the light of his monitoring of his own message-reception process, concludes that the categories being used to accept information are either too simple or too complex for the message which is apparently available, and takes a decision to adjust them [18].

When such links between information and order are also associated with the hierarchical structuring of energy inputs, we find that we are very close to the processes by which living organisms turn energy from the world to their own advantage, to build up their own internal and external order processes in which information has an important status in the sustenance of life itself. Information thus has, it has been suggested, an 'ontic status'. [19].

8.3. *Unexpected events*

The relevance of this extended discussion of aspects of information theory to disasters and large-scale accidents should begin to become clear when it is realized that all such incidents are unexpected to some degree, and that they often constitute the kind of event which leads to a change in the interpretation of the environment, and to a change in the kinds of 'messages' which the environment may be thought of as 'sending' or making available to us.

We have seen in the foregoing discussion that surprise at an unexpected event is not necessarily always surprise of the same kind: events which are surprising but uncertainty-reducing, when they occur, provoke what we might regard as 'normal' surprise. Events which increase our uncertainty initially, because we have no immediately available framework for handling them, provoke a higher order surprise which is associated with a discontinuity in the information-acquisition and uncertainty-reduction process.

The kinds of event which may provoke a higher order of surprise can be separated into three groups—which we may label as anomalies, serendipities and catastrophes. All three types of event share the common property that the news of their occurrence does not reduce uncertainty, or at least, does not do so immediately. *Anomalies* are pieces of information which are clearly not irrelevant to the concerns of those who receives them, but which cannot be assimilated into the existing world-view, so that their implications for understanding and for decision-making cannot be fully assessed at the time of their acquisition. The term 'anomaly' suggests links with Kuhn's discussions [20] of the way in which scientists come to accept new theories or paradigms about the world, and this association is a conscious one, for the problems faced by both administrators and scientists, in building up and maintaining a coherent picture of the world within which they are trying to operate, display a number of similarities in the context of this discussion.

In a scientific context, the odd markings on the radio-telescope scan which provided the first indication of a new type of 'pulsar' star to the Cambridge radio-astronomy team was, at first, an unclassifiable piece of information which initially served to increase rather than to decrease uncertainty [21]. In an administrative context, a section of the Ministry of Transport which compiled routes for abnormal loads reported to the public inquiry following the Hixon level-crossing accident that, when they were told of the existence of a new type of automatic barrier crossing at Hixon and at other places, they had spent some time considering the question of 'exactly what category of information this particular feature came into', without relating it in any satisfactory way to the assumptions and categories with which they operated when devising routes for abnormal loads [22].

Both in science and in administration, anomalies are noted as curiosities, as events which are not readily assimilable, but also as events which there is no immediate and pressing urgency about elucidating. Anomalies may be labelled and shelved, to wait until a superior explanation comes along to include them, and sometimes their existence 'flags up', to an inquiring mind, the possibility that there may be data here which could serve as a starting point in a search for a new level of information [23].

The other two forms of unexpected event which increase rather than decrease uncertainty may be dealt with together; for the difference between what might be called a 'serendipity' [24] and a 'catastrophe' is the difference between an unexpectedly favourable and an unexpectedly unfavourable precipitating incident, and for present purposes this difference may be assumed to be attributable to random factors. In both types of case, major pieces of information are discovered in unexpected areas, with implications for the accepted view of the world, and the action-related assumptions flowing from it, so that a revision of explanations previously accepted as satisfactory becomes necessary. In both cases, the consequences draw attention to the discrepancy between the view of the world enshrined in the relevant sets of premises upon which decisions are based, and the additional external factors which now have to be recognized. Much more data is available about catastrophes than about serendipities, however, for 'unexpectedly lucky breaks' do not usually create problems for action, and there is not normally a public inquiry to challenge the natural claims that any administrators or managers concerned are likely to make: that they really had planned it that way all along [25].

As we have seen before, accidents and disasters always arise as a result of some form of discrepancy between the way in which the world is believed to be and the way it really is. Even when an activity is recognized to be dangerous, those who undertake it without a suicidal intent have some belief that they will cope, that they will manage to avoid an accident. And, because of the importance of organizations in our society, discrepancies in official administrative views of the world which lead to accidents are particularly significant. Small discrepancies in the views of the world shared by groups of managers and administrators within organizations are constantly accommodated to and adjusted to by means of minor modifications and revisions of assumptions, and by alterations in the practices associated with them. But when events with the informational properties associated with 'anomalies', 'catastrophes' and 'serendipities' arise, more radical shifts of assumptions and practices become necessary, if the organizations are to continue to appear to conform to the norms of rationality. As a development of the definition of disasters set out in Chapter 5, therefore, we may suggest that it is useful to classify accidents, disasters and other events of this kind, not necessarily according to the severity of the consequences which follow from them, but according to their unexpectedness with regard to the prevailing institutionally accepted models of the world, as indicated both by the number of revisions of administrative practice which are seen to be necessary after the incident, and by the extent of such revision [26].

8.4. *Surprise and the differential awareness of hazard*

Surprise is an essential ingredient of accidents and disasters, but surprise is not always a simple matter. For any given unanticipated incident, there will be a whole array of reactions, both cognitive and emotional, from different individuals according to their closeness to the incident and the extent of their personal involvement. If we set aside the differing emotional responses, and concentrate for a moment upon the cognitive reactions, we can see fairly readily that different individuals will be surprised to different degrees by the same incident, to the extent that their prior expectations about likely events differed.

Knowledge about hazards, about precautions, about likely safety levels and about the probabilities of different kinds of accident is not evenly spread throughout society. Instead, such knowledge is distributed differentially between different individuals and groups; and, in a similar way, the ability to take 'reasonable' or 'well-informed' decisions about the level of risk to be tolerated is also held unevenly by different groups and individuals.

The analyses of accidents and disasters presented earlier suggest that, with respect to any given set of dangers, there is likely to be one major institutional group, or sometimes two or more groups, in particularly important positions in that they are able to influence the way in which these dangers are seen and understood by others. Powerful groups and organizations are able to specify the kinds of hazard that they recognize, to set out and implement the kinds of precautions which they think are necessary, and to exert their authority in intervening in areas which they regard as hazardous. But these bodies are not the only ones involved, and, at times, their understanding and their definition of the hazard may be challenged.

There may be confrontations between those who present an official definition of hazards and others who think that the situation is different, which is to say, in virtually all cases, more dangerous. But confrontation is not always necessary, for it is quite possible for pressure groups to be active in lobbying and exerting pressure over issues of this kind, and for the institutionally accepted view to be modified as a result [27]. The manner in which hazards are perceived will also vary to some extent according to the vested interests which those concerned have in different aspects of the dangerous situation; for example, nineteenth-century factory owners would have seen the hazards of factory work and the precautions which might be regarded as 'reasonable' very differently from those working for them. And, equally, at times, this situation may be reversed, when incentives are in operation which lead employees to ignore prescribed safety procedures. There is thus an overlay of differential power distributions which will affect knowledge, perceptions and expectations of accidents.

Having made it clear, then, in a general manner, that knowledge and expectations with regard to a range of hazards will be unevenly distributed amongst the organizations and individuals in our society, we can see that the most unexpected events will overturn not one, but many differing sets of expectations when they occur. And if any one individual or group is virtually *certain* that a particular danger will produce a calamity within a specific time period, we have one individual who will *not* be surprised when the accident occurs. He may well feel that he should use his knowledge to warn others of the danger, as, for example, a man knowing of an obstruction on a railway line would try to warn the driver of an oncoming train of the danger ahead. But even this situation is not quite that simple, for there is always the possibility that such a warning may be wrong, and the individual concerned will need to be able to convince the other parties involved that his warning is a genuine one, by producing evidence which meets their minimum criteria for validity.

A very precise warning does force those concerned to take a decision about whether the warning is genuine or not, but in most pre-accident situations, matters are much less cut-and-dried. More commonly, the differences between parties about possible future dangers will be much vaguer with regard to the likely timing, the likely location of the danger and the precise form of the danger with which they are being confronted. On any or all of these issues, the differences of opinion may well be of the order of disagreements between, say, the '5% probables' and the '25% probables'. The inhabitants of Aberfan were apprehensive in a general way about the tips overhanging their village, but they would have kept their children away from school if they had been able to predict with considerable accuracy when and where the tips would slide. Their apprehension, however, allows us to speculate that they were less 'surprised', in some sense, than say the Chairman of the National Coal Board, who appeared to have had no such apprehensions about tips in general, or about the Aberfan tips in particular.

When we talk, then, about discontinuities and surprise evinced by unexpected events, it is clear that we are generally referring in an approximate manner to an aggregate response of different levels of surprise encountered by many different groups and parties [28]. Bearing this qualification in mind, we may turn to look at another aspect of the nature of surprise and discontinuous information-acquisition, before returning in the following chapter to a more detailed discussion of some actual instances of accidents encountered in administrative and organizational settings.

8.5. *Catastrophe theory*

Much of the discussion already set out in this chapter has been concerned with the discontinuity which arises when a channel of observation ceases, momentarily, to operate as though it were a channel of communication. Until recently, there has been no way in which our understanding of such abrupt transitions could be aided by the use of mathematical tools, but with the publication of Thom's theory of mathematical 'catastrophes', a body of work now exists which makes it possible to apply mathematical analyses to discontinuous as well as to continuous changes [29]. It should be noted, since we are talking about Thom's work in connection with disasters, that his catastrophes have nothing to do directly with the kinds of phenomenon that we are interested in: he seems to have used the term 'catastrophe' to apply to abrupt changes in behaviour, particularly in the organic field, as when a cell suddenly divides into two. There are indirect links between disasters and Thom's notion of a mathematical theory to account for sudden organic changes, however, so that it is worth looking at the way in which Thom's discussion of discontinuities may be related to our examination of accidents and disasters.

The mathematics involved in developing Thom's theory is of a most complex and advanced kind [30] but, fortunately, some of the outcomes of it lend themselves to presentation in a fairly simple and readily comprehensible form. Without pretending to grasp fully the mathematics involved, we may take over from Thom a very simple way of representing discontinuities in graphical form, which extends the analysis already presented above. Figure 8.6 represents a section through one of the simpler of the three-dimensional 'catastrophe' functions which Thom has identified, a 'fold' catastrophe. As a point travels along the curve from A to B, it enters a zone, C—D, where any value on the x axis has three possible values on the y axis. Inertial forces of some kind, however, keep the point on the lower curve until the point X is reached. Any further movement causes a discontinuous leap to Y, and the point then continues its smooth journey to its destination, B. Since a similar leap, Z—Z', would be made by a point travelling in the reverse direction, it is clear that the intermediate section of the curve, S, is essentially inaccessible, so that within the range C—D we are normally concerned with two and not three possible positions on the curve.

If we try to relate this kind of curve to our earlier discussions of discontinuities, we may plot uncertainty against time, and begin at A (Figure 8.7), uncertainty being progressively reduced as more information is acquired with time. At point C we enter a zone in which two levels of uncertainty are possible, a possibility created by the discrepancy between intention and actuality. Perceptual and institutional inertia en-

INFORMATION, SURPRISE AND DISASTER

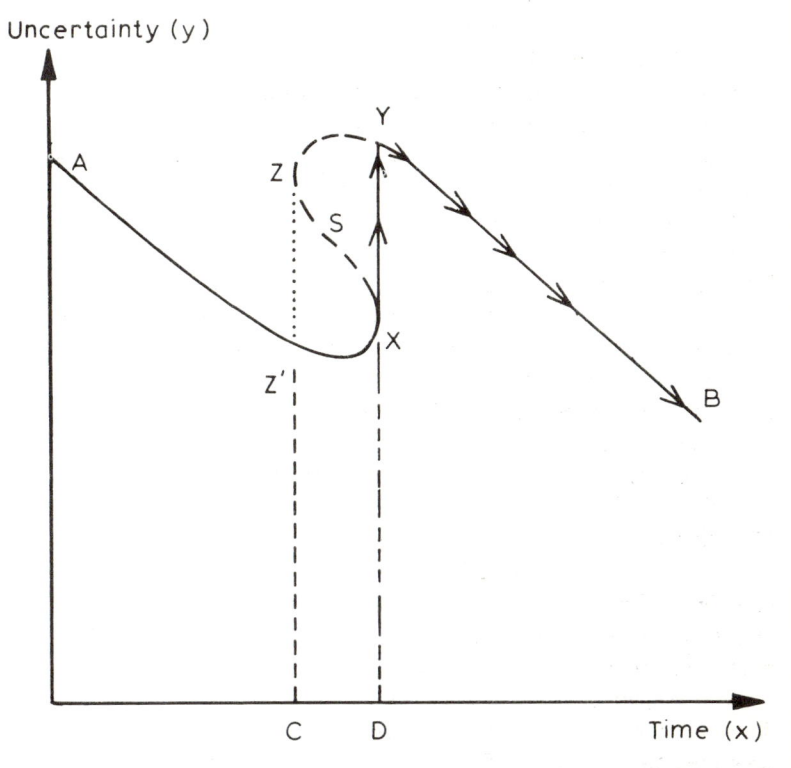

Fig. 8.6. Representation of a section through a 'fold catastrophe'.

sures however that, at this stage, information-accumulation continues to be assessed in the light of the initial framework of expectations (I), within which information had been gathered. This framework defines a possible set of messages which could be accepted and, within the framework, conventional communication theory operates as an adequate means of describing the effects of successive information-acquisition events. But when the last of a series of information inputs occurs, to bring us to point X on the curve, any additional information relates to a 'precipitating incident', and provokes a discontinuous, rapid leap from the previous context or framework of understanding (I) to a new framework or context (J), with a new set of possible messages associated with it (Jm)—see Figure 8.7.

Anomalies in this kind of diagrammatic outline would occur within the region C—D, hinting at the existence of another context, but not challenging the existing one sufficiently strongly to precipitate a change

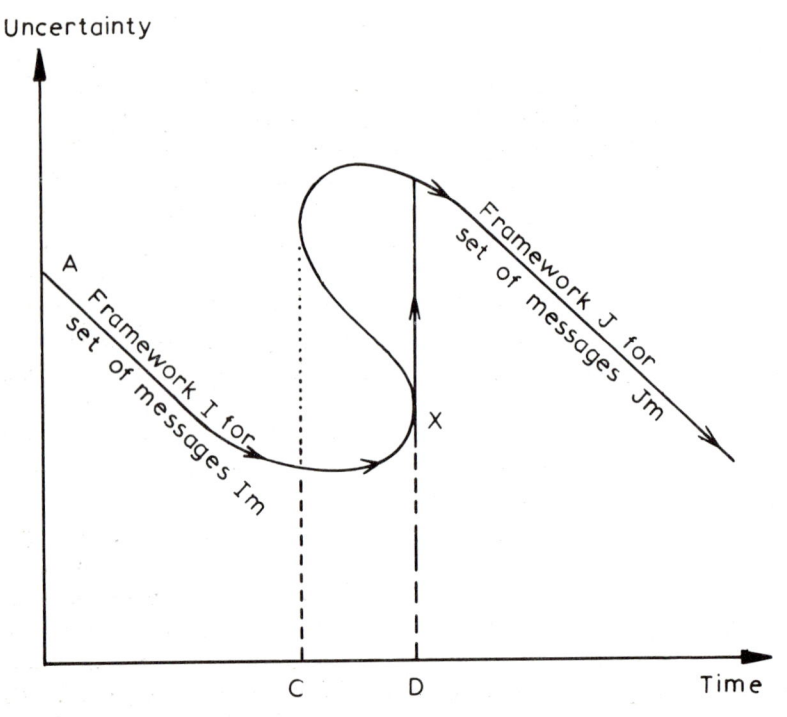

Note: The process of acquiring information is an inherently discontinuous one, so that the curve presented above must be seen as an approximation. However, as we move from the simple case of a two-state framework towards more complex frameworks, the approximation becomes closer.

Fig. 8.7. Framework shifts represented as a fold catastrophe.

of context. The kind of precipitating event which does produce a change of frame, or a shift in the basis for understanding, is provided by what we have already referred to as a 'catastrophe' or as a 'serendipity'. Alternatively, the jump could arise as a result of a new understanding generated from an accumulation of anomalies. Figure 8.7 represents this discontinuous shift by plotting information against time, but Figures 8.4 and 8.5 have shown that it is also possible to represent the same shift by plotting uncertainty against information, a plot which yields a simple straight line for communication channel applications, since information here is, by definition, a measure of the increase in certainty.

This kind of discontinuous process has been noted in a number of settings by various writers: there are evident parallels, which we have already alluded to, between the jump described above and the contentious notion of 'paradigm shift' which is found in Kuhn's discussion [31] of the progress of scientific thought. This kind of discontinuous process has also been referred to in social and psychological contexts. The anthropologist and psychiatrist Gregory Bateson has pointed to the existence of discontinuous jumps which arise in the learning process as a result of moves up and down a 'hierarchy of learning' [32]. At the bottom of this hierarchy is a state of 'zero learning', in which there is no adjustment of future behaviour as a result of past information-acquisition. Above this, Bateson locates what he calls 'Learning I' which is learning in a situation with a given set of alternatives, in which one alternative rather than another comes to be preferred as a result of experience. Above this again is 'Learning II', as a result of which the individual comes to make a corrective change in the sets of alternatives from which choice can be made. The parallels between such an approach to learning and the discussion earlier in this chapter, of the differences between channels of communication and channels of observation, is very clear.

Bateson, in commenting upon this process of discontinuous shifts, observes that for each stimulus or information input, the context or framework within which it is received is a meta-message which classifies the elementary signal, and this observation leads him on to the idea of 'context markers' which indicate when one context rather than another is appropriate. The sociologist Erving Goffman has taken up and elaborated this idea of context markers in his recent theory of frame analysis, applying the idea of discontinuous shifts up and down a hierarchy to aspects of the whole range of face-to-face social interactions, demonstrating that such shifts are basic to human interaction, and that we are highly skilled perceivers and users of such shifts [33]. Other research [34] has suggested very strongly that human beings find the condition of 'not having an acceptable frame to operate in at all' to be intolerable; so that it is difficult to disrupt existing frames and channels of communication, which are clung to tenaciously and with great ingenuity, and when disruption does take place, a rapid context shift also takes place, a new context or channel of communication being set up with great speed.

The existence of these observations and theories relating to discontinuous jumps in perception and information-acquisition in a range of fields of study is interesting. We have identified a pattern of information-acquisition which seems to be important in understanding human responses to unexpected events such as accidents or disasters. The transition in understanding provoked by such events, the transition which leads them to be classified in retrospect as 'accidental', is a phenomenon which is related to certain basic characteristics of human intelligence and human response to the world which have been observed by others. The crucial common property to which we are referring derives from the realization that information-acquisition, understanding, learning and surprise are all ordered, not on a single level but at a number of levels, so that progress from one level to another takes place by means of discontinuous jumps. The American psychologist John Platt [35] has labelled this process 'hierarchical restructuring', a process by which the emergence of order at a higher level occurs, through hierarchical jumps, as new information is brought in. He suggests that such jumps have four characteristics:

1. The jumps are preceded by and accompanied by 'cognitive dissonance'—or by what we could, perhaps, refer to as an increased state of uncertainty provoked by an awareness of the existence of anomalies;
2. Both the dissonance and the jumps have an overall character;
3. The restructuring is sudden when it occurs;
4. The new structure offers a conceptually simpler and more general way of organizing available information than the old structure.

Platt mentions physical, biological and social examples of discontinuous hierarchical restructuring, and draws attention to some interesting suggestions that the most common pattern of restructuring is one which enables the major elements of the system prior to restructuring to retain some of their relationships throughout the transformation. These elements of the old system thus persist into the new one, in a changed manner, with necessary adjustments being made in peripheral elements of the system [36]. When a jacket is turned inside-out, the sleeves, pockets and collar remain in a very similar relationship to each other both before and after the transformation, although the overall effect is quite different; in an analogous manner, our understanding of a situation before and after a discontinuous hierarchical shift retains most of the elements of our initial beliefs and perception, but is transformed because a new interpretation of the old pattern is now possible. The diagrams in Chapter 5 relating to the Cambrian Colliery accident illustrate the manner in which such a shift may occur in connection with accidents and disasters, the new interpretation of the situation emerging even though many features of the old interpretation remain.

8.6. *Summary*

This chapter has introduced the notion of information as defined in communication theory, and has considered measures of information relating to reductions in uncertainty. Outside the closed-system model of the communication channel, however, the usefulness of conventional measures of information diminishes; for there is a need to take account of subjective elements contributing to the uncertainty of the receiver, of the difficulty of classifying messages for transmission, and of the reinterpretation of messages which may occur when the monitoring of transmissions provokes a discontinuous shift in the manner in which information is interpreted. These characteristics can be summarized by referring to 'channels of observation', which constitute open systems, rather than to 'channels of communication'.

Discontinuous jumps of this kind may be provoked by various types of information input which are monitored: by anomalies, by serendipities and by catastrophes. Individuals will be differentially surprised by unexpected events; the discontinuities may be characterized by means of the use of catastrophe theory, and such discontinuities seem to have a wide application in social, psychological and other fields.

9. Disasters and rationality in organizations

After the rather abstract discussions of the previous chapter, we can return now to an examination of some of the social and organizational contexts within which the conditions creating disasters may develop. It is only very rarely that the emergence of a disaster can be fully attributed to the blunders, errors or misunderstandings of a single individual. In most cases, as the examples reviewed earlier have suggested, it is necessary to take into account not only the contributory behaviour of several individuals, but also the manner in which the behaviour of these individuals is shaped by the institutions and organizations within which they act. Thus, in discussing the errors and communication difficulties preceding a number of disasters in Chapter 6, we found it necessary to relate the communication patterns under review to the organizational settings within which they were located, and, at a slightly more general level in Chapter 7, we began to see some of the interrelationships between individual and organizational decision-making, and to understand the way in which limitations upon intended rationality at the individual and at the organizational level are linked.

The part played by various organizational processes in the development of disasters will be further examined in this chapter, as a central aspect of our inquiry. With the preponderance of organizations in the modern world it is likely that, in virtually every case of disaster, there will be an organizational involvement. Individuals are rarely in a position to create major disasters without official or unofficial access to the energy and other forms of resources controlled by organizations, for organizations have a near-monopoly of control over access to most of the sources of energy which could be discharged to produce disasters. And, because organizations are so closely involved with the forces and conditions likely to produce disasters, it is inevitable that anyone trying to understand the conditions likely to produce disaster will need to examine the manner in which organizations handle information about potentially dangerous energy sources of all kinds, and to examine the way in which errors in the handling of this information develop and are transmitted. This means that it is necessary to try to gain a better understanding of the manner in which the constraints of bounded rationality operate within organizations. In the course of the discussion, it should become clear that there are important parallels to be discerned between, on the one hand, the hierarchy of decision-making assumptions which characterize organizations, and, on the other, the hierarchy of information levels discussed in the last chapter. The misunderstandings and misinterpretations which enable disasters to develop emerge in various ways from the flow of communications within organizations, between organizations and between the general public and organizations. As organizations carry out their normal day-to-day business of decision-making, they draw upon and sustain an official model, an official understanding of that environment which they are trying to influence. When this official understanding is grossly in error, the potential for a major disaster exists, so that it is important to examine the manner in which organizations create and maintain the sets of assumptions upon which their decision-making rests. We shall therefore draw together some of the issues already touched upon in earlier chapters in order to try to understand the manner in which failures of foresight may develop within organizations [1].

9.1. *Rationality in organizations*

It has already been pointed out that our society places a particular value upon rational behaviour, upon behaviour which forms part of an extensive, predictable and consistent system of actions carried out in order to achieve some specified end. This kind of behaviour is a particularly elaborate and controlled response to man's urge to behave purposively, and large organizations in Western society are institutions which embody this kind of response within their structure. The ideals of rational behaviour are believed to provide some form of standard against which behaviour can be measured, and rational patterns of response grow and develop as a result of the examination and criticism of previous responses, undertaken with the aim of improving the suitability and effectiveness of actions to be taken in the future.

When an individual or an organization seeks to attain a goal within a completely closed system, it is possible in principle for a complete knowledge of the system and of its options and probabilities to be obtained, so that a completely rational pattern of response can be drawn up. Rational strategies for solving certain restricted kinds of puzzle, for example, or for determining, say, the chemical composition of a given substance, are thus readily available. In such cases, since the limits of the problem are known, the actions of those tackling the problem may be reliably judged against the standards set by rational norms.

In most real-life situations, however, and particularly in the face of the most important problems to which decision-makers are required to address themselves, the issue is much less simple; for, in such cases, they are usually dealing with open systems in which they are unable to gain a complete knowledge of the implications of one action as set against those of another. In these situations it is widely accepted that the most rational response lies in taking the best decision possible, given the information available. Where the system approximates to a closed system, it is reasonable to present the decision-maker with an array of probabilities or likely frequencies of outcomes which will follow his choice of a given alternative; and techniques are available to enable him to formulate 'best' or most rational strategies in such circumstances, even though it is recognized that his choice may be invalidated by subsequent events which he did not anticipate [2].

In more complex real-life situations, where frequencies and probabilities are not readily estimated, decision-makers still try to approximate to the kind of behaviour which can be handled by these techniques; so that if they are attempting to be particularly rational, they may set out arrays of possible strategies and eventualities, and attempt to assess their choice of the 'best' strategy, given certain assumptions, certain pieces of information, and certain areas of ignorance. The systematization provided by such a disciplined ordering of the issues may be helpful in many cases, but where the number of imponderables is great, all that may result is the cloaking of ignorance with a layer of false precision [3].

But if we are looking back upon a decision which has been taken, as are most decisions, in the absence of complete information, it is important that we should not assess the actions of decision-makers too harshly in the light of the knowledge which hindsight gives us. If we are to extend our understanding of the way in which erroneous decisions lead to accidents and disasters, we need to assess any given error in two ways, at two rather different levels.

Firstly, we need to determine whether the decision taken was a 'reasonable' or a 'rational' one, given the information available to the decision-maker at the time. If it can be demonstrated in such an assessment that an alternative choice ought to have been made on the basis of the information then available, we could be justified in characterizing the decisions as 'incompetent', or as indicating 'poor judgement'. But it is necessary to be extremely careful when making such retrospective judgements, for experiments have clearly demonstrated that assessments of likely outcomes by individuals *after* the event are markedly affected by their knowledge of what the actual outcomes were. Moreover, the same investigations have shown that people are likely to be unaware of the manner in which their assessment is shifted by such knowledge [4].

The second kind of assessment of erroneous decisions which we might make becomes necessary if we have concluded that the decision-maker did act reasonably, so that other competent persons faced with the same choice in the same circumstances would have been likely to have made the same decision. In such circumstances, it is clear that the source of error is not to be found in the behaviour of the individual, who has conformed to the norms of rational or intendedly rational behaviour. What then needs to be assessed is the question of how a situation could come about in which reasonable men, attempting to behave rationally, could still be in error [5]. To answer this question, it is necessary to look at the manner in which rationality becomes established and embedded within organizational procedures and habits, and to gain an understanding of the hierarchy of decision-making within which the individual administrator finds that he has to operate.

9.2. Co-ordinating individuals within organizations

This issue really goes to the very heart of the processes which make modern organizations possible, for if we are to understand how an organization can attempt to act rationally in the world by combining and co-ordinating the separate judgements and intellects of all of the people who make up the organization, we have to understand more generally how an organization works. One of the problems which organizations repeatedly have to face in order to ensure their continued, effective existence is the problem of integration. The essential ingredient of any organization is that it should have some capacity to act as a single entity, to act in an integrated manner, for if it does not have this capacity it ceases to exist. But two factors hinder integration. The first is the diversity of the individuals who come together to make up the organization, bringing with them their own beliefs, hopes, fears, rationales and idiosyncrasies. The second factor is the element of necessary diversity which the organization must develop and contain within itself in order to cope with the complexities of the world within which it operates.

Every organization, therefore, must seek to establish some internal discipline, some internal set of constraints to ensure that its members do act as a unit. At the same time, it must retain the valuable individual capabilities for which its employees were recruited, and maintain specialist sub-units with the specialist knowledge of the world necessary for the organization's survival. The factors, both internal and external, which encourage diversity, will continue almost independently of the organization. The organization needs, therefore, to oppose these tendencies without eliminating them altogether, and it does so by means of three interrelated processes: it establishes a hierarchy of authority and decision-making which enables different problems or sub-problems to be considered at different levels of generality within the organization; it permits those with the greater power within the organization to operate through the formal hierarchy and in other ways, to encourage actions, intentions and analyses which they approve of and to discourage conflicting views; and it ensures a continuity of outlook, practice and knowledge within the organization by means of the social processes of communication and socialization which enable the organization to build up and maintain a sub-culture embodying the organization's distinctive character and mode of operation.

Having pointed out in this very general way, the crucial importance of hierarchy, of power and authority and of beliefs and culture within organizations, we may now go on to consider in rather more detail some of the ways in which organizations strive to achieve an integrated form of unified, rational action. Subsequently, by means of the examination of some instances of accidents, we shall try to understand what happens when organizations fail to achieve the goal of rational action which they are pursuing.

We must ask, first of all, how the situation of an individual changes when he becomes part of an organization. We have already seen that, for the individual, the goal of 'pure' or 'global' rationality, permitting the choice of the best possible alternative, is unattainable except with regard to trivial decisions. Because of finite limits upon the individual decision-maker, the best that he can hope for is that he 'satisfices', that he chooses an alternative leading to a satisfactory outcome. His potential for rational behaviour is always 'bounded' by such limitations.

How does the above argument apply, then, when a person finds himself acting within an administrative context rather than as an isolated individual? His capacity for rational decision-making will of course, in most cases, be extended by access to those resources of the organization devoted to information-gathering and information-processing; so that the executive or manager within an organization is likely to be in a better position to cope with the uncertainties and complexities of the environment than he would be on his own. But even so, the resources of organizations are not infinite, and, when we take into account the larger scale of the tasks that organizations tackle, we soon realize that there are bounds set, too, to the capacities of collectivities of decision-makers. Some of these bounds are set because organizations cannot afford to extend their information-processing capabilities indefinitely, but other limits exist because of the organization's need to maintain that minimum degree of integration upon which coherent corporate action depends.

To understand more of these limitations upon the potentialities of the decision-maker within organizations—of 'administrative man' as Herbert Simon calls him—we need to look in a little more detail at the precise nature of the linkages between decision-making at the individual and at the organizational levels, and at the ways in which these linkages might contribute to the incidence of gross errors [6].

We can think of the decision-maker, on taking up his post within an organization, as entering into some kind of 'psychological contract' with the organization which leads him to accept, to a greater or lesser extent, the organization's definition of the role which he is expected to play.

The organizational role which he accepts will offer to him, or will impose upon him, a set of assumptions, criteria and premises which will serve to guide his decision-making by informing him of the kinds of criteria which are acceptable as rationales for decision-choices within the organization concerned [7].

The constraints which these assumptions and decision-premises impose upon the individual in his official capacity as a decision-maker within the organization produce an interaction between the bounded rationality of the individual when he is acting in his official capacity and the much more extensive, but still limited, rationality of the organization as a collectivity. By specifying, to some extent, the decision-premises which the individual accepts as part of the role which he negotiates with the organization, it is possible for the organization to exert an influence over the whole paraphernalia of sets of alternatives, perceived and unperceived, and the sets of outcomes and pay-offs which are to be considered when any issue is being debated.

At an overt level, the organization can be thought of as providing a list of pay-offs which are thought to be important to the organization, and more or less well developed rules for choosing between options which promise these pay-offs. It may well also specify a form of rhetoric in which choices must be outlined and defended if they are to be seen as legitimate within the organization [8]. But at a deeper and less explicit level, the assumptions which constitute decision-premises convey to the administrator a basis for the construction of his interpretation of an officially approved 'world view' in terms of which the organization sees its relationships with the world.

Without these decision-premises, without this world-view in at least a rudimentary form, an organization would cease to have any integrity of action. Decisions taken by different individuals would seek to move the organization in differing directions, creating the danger that the organization might disintegrate into nothing more than a crowd of separate people with different purposes and different ideas about the attainment of these purposes. But the constraints of a more or less unitary set of decision-premises applied to the individuals within an organization enable that organization to retain its singleness of purpose by ensuring that all of its key decision-makers are operating with variants of the same organizationally approved bounded rationality. As a caricature, it could be said that organizations achieve a minimal level of co-ordination by persuading their decision-makers to agree that they will all neglect the same kinds of consideration when they make decisions!

But, in drawing attention to this integrating force which holds the actions of any organization together with at least a minimum amount of coherence, it must not be thought that we are overemphasizing the singleness of purpose to be found in most organizations. Rather, this element of integration must be seen as one factor within a complex diversity of elements which make up the processes found within an organization. The assumptions and beliefs which constitute decision-premises will form just one part of the culture or sub-culture which exists within an organization, and they will be transmitted and stored, preserved and modified in the same way that other elements of the organizational culture are. Information, views of the world, norms and beliefs about appropriate behaviour vary from organization to organization, from department to department and from office to office [9]. There will be competition and conflict between different elements of these organizational beliefs, and some writers have suggested that the more successful organizations are those that are able to bring conflicts and disagreements about issues relating to their central tasks out into the open and thus to resolve them [10].

Even where such overt resolution of differences does not occur, it can be seen that every organization faces the problem of producing a degree of unity from the diversity of its individual elements, and this orderly collective response is produced, as we have already suggested, by three mechanisms. The sharing of tacit assumptions about the world is achieved to some degree by the processes of socialization and culture-formation which go on within any social grouping. These processes, in an organization of any size, however, still produce many particular world-views and many sets of idiosyncratic perceptions, and the internal operation of forces of power and influence then dictate which of these 'micro-cultures' shall be the dominant ones within the organization [11]. The views of reality, which are most likely to be dominant, and the decision-premises which are most likely to be dominant are the ones favoured by 'those with the bigger stick' [12]. The selection of the most dominant elements from the diverse sets of views available within an organization will also be aided to some degree by the hierarchical manner in which authority is distributed within organizations and by the parallel hierarchy of sets of premises and assumptions.

Administrative man, then, the decision-maker within the organization, has available to him, from his superiors or his colleagues, decision-premises which are associated with his position in the organization, and with the 'micro-culture' within which he finds himself. In the course of his normal activities, as he learns about his job, he will continually be making exploratory trial decisions and monitoring their outcomes, drawing upon accepted knowledge about previous outcomes and upon the 'micro-cultural' judgements of his immediate colleagues as to the appropriateness of different strategies. Even if he is not particularly innovatory in his activities, each time that he implements an established procedure, he confirms its efficacy and adds another instance to the set of case-law available to him and his colleagues. By such means, the individual contributes to or modifies the existing stock of decision-premises. If past experiences are noted and remembered by members of the organization, they become part of the local organizational memory, the local organizational experience and wisdom, available for application to future cases. Under stable conditions, the operation of such processes serves to establish firmly the terms, limits, constraints and areas of operation of a given set of decision-premises; while under conditions of change, by means of very similar processes, gradual modifications may be brought about in decision-premises as additional fragments of information are added, over time. When the premises upon which decisions have been based are grossly in error, however, so that an accident or disaster results, there is pressure for a much more abrupt shift in organizational assumptions.

Using this idea of decision-premises, it is possible to shift our investigation of errors arising as a result of the limits of bounded rationality away from the situation in which a single actor is trying to behave rationally to situations in which *groups* of decision-makers are trying to behave rationally, judging their success by means of criteria which are approved within their organization. When several organizations are engaged in tackling interrelated issues, and when members of the local or national community are involved in the decision-making process, it may also be appropriate to think of these additional groups of decision-makers as operating with their own sets of decision-premises.

It seems to be worth reiterating, however, that this idea of decision-premises should not be over-simplified, even though some of the comments above might be interpreted as suggesting that members of a given organization will all have imposed upon them, as the price of their membership, a strait-jacket which ensures that their thoughts and decisions move only along predetermined lines. The true situation seems to be both more complex and more subtle than that. It is more subtle in the sense that the sets of assumptions upon which it is expected that individuals will assess a situation and make a decision will never be clearly or openly specified in full. The individual has to judge or to find out by trial and error what is reasonable and acceptable, and he has the opportunity, too, to challenge and modify some of the prevalent assumptions if he has the necessary ability and influence. These kinds of changes enhance the complexity, for while an organization cannot be expected to cohere unless there is some degree of agreement expressed in the official actions and official decisions of a significant number of its members, the precise assumptions about what may be regarded as 'reasonable' and 'rational' will vary from department to department within the organization, and they will also be subject to modification over time. For example, a sales department, a production department and a research department will be oriented to different sections of the environment, and each will have to develop an ethos based upon its own sets of assumptions about the environment if it is to perform its task. The organization of which these departments form the constituent parts will then have to cope with the problem of balancing this differentiation, which each department needs in its relationship with the world, with the need to integrate all of these contributions in the pursuit of goals which are seen as desirable and worthwhile for the organization as a whole [13].

9.3. *Hierarchies of decision-premises: Bounded decision zones*

Amongst any group of administrators, to the extent that the cultural pressures within their organization make their outlooks on organizational matters homogeneous, there will be a shared perception of the events likely to affect issues important to them, and of the probable outcomes of given actions and decisions. Consequently, within any organization, we may expect to find many groups of decision-makers, each with their own assessments of alternatives, of outcomes, of pay-offs, and of the calculus linking these elements; so that there will be a number of zones within which slightly differing decision-premises will operate. Each such zone (and there will be, of course, overlapping clusters of zones) will thus be associated with a world-view congruent with the set of decision-premises accepted within it.

We may try to state this expectation a little more formally, by describing a *bounded decision zone* as one which approximately but effectively, for those within it, embraces all officially admissible perceived possibilities which the individuals are prepared to consider in their role as members of the organization. That is to say, the sum of the uncertainties which relate to matters of organizational operation, which are not negligible in day-to-day decision-making terms, and which lie outside the bounded decision zone are, for the members of that zone, approximately but effectively zero [14].

To illustrate this point with a simplified example, we could imagine a group of sales managers who, in considering their future policy, might behave as though they assigned the following probabilities to the likely movement of their product in the market during the coming year:

Probability of sharp increase in sales:	0·2
Probability of steady increase in sales:	0·5
Probability of little change in sales:	0·2
Probability of slight fall in sales:	0·1
Total:	1·0

The sum of these perceived possibilities is unity, and if the planning of these managers is conducted on the basis of such assumptions, whether they are made explicit or not, it is clear that there is no room in such a scheme for eventualities such as the company being taken over and closed down, the product being entirely superseded or nuclear war being started. The probability assigned to these eventualities we may assume to be effectively zero. A bounded decision zone is thus a zone within which there is a shared bounded rationality, defined by a common set of perceptions and decision-premises which define a field of likely, and therefore to some degree anticipated, events. When one rather than another of the expected events materializes, the administrators concerned experience a collective surprise of what we called previously a lower order, but when their officially enshrined expectations are overturned, we may regard those within the bounded decision zone concerned as experiencing surprise of a higher order.

The nature of the bounded rationality being discussed here includes only those possibilities which are taken into account in practice in the decision-making process, for this then makes it possible to note the difference between views which may be endorsed officially within the organization, and views held privately by members of the organization. While an individual official or manager may not be surprised if the product on which he is working turns out to be a 'dud', or if the Second Coming interrupts stock market dealings next Wednesday, unless his anticipation of these events has been registered in the bounded rationality of the organization, the organization as a whole will be surprised, even though some of the individuals within it are not [15].

Taking the idea of bounded decision zones one step further, we can ask how we might apply this idea to a hierarchical set of groups of individuals, concerned at their different levels with taking decisions about interlinked but not identical areas of concern. Such a hierarchy, which is to be found within any modern organization, may be thought of as hierarchy of overlapping bounded decision zones, which has associated with it a hierarchical set of bounded rationalities, each defining rules for describing a set of informational boxes within which the organization is prepared to receive messages about its environment [16]. With the image of such an information-receiving and processing hierarchy in mind, we may, perhaps, be equipped to gain an understanding of some of the ways in which organizations cope with the world, some of the ways in which they adjust to changing circumstances in the environment, and some of the ways in which they react when they discover that the assumptions of one or many of the bounded decision zones within their hierarchy have been grossly in error: to examine, that is, the response of organizations to accidents, disasters and other failures of foresight.

To recapitulate, we have suggested that in order to understand the nature of bounded rationality within organizations, particularly with regard to its limitations and failures, it is necessary to take account of the manner in which sets of decision-premises are distributed within organizations, and to take account of the way in which this social location of sets of decision-premises influences the actions of the organization. Sets of decision premises may be thought of as being arranged in an overlapping hierarchy of zones which are internally homogeneous in terms of their official world view. Individuals related to various of these zones assess their reasonableness or rationality by drawing upon the view of the world which is available and at least partially approved of within their portion of the organization, and by using the set of premises associated with this view. Each individual may then, according to his own idiosyncrasies and according to his official position, create a different variation upon this central theme.

DISASTERS AND RATIONALITY IN ORGANIZATIONS 171

Within the interlocking and partially overlapping micro-cultures associated with the various decision zones, because of the different premises and the different information available to groups, pressures for separation arise. But, acting against the effects of this differential distribution of information and interests, there are also opposing pressures which work to keep the organization together. These pressures, deriving both from the influence of the more powerful groups in the organizational hierarchy and from the effects of the cultural forces within the organization, contribute to the production of those recognizable similarities in the way in which organization members act and react which are referred to by such labels as 'organizational style'.

9.4. *Some examples*

Having set out at some length a number of ideas which enable us to grasp the kinds of process which link individual and collective rationalities within organizations, we may now use the understanding offered by these ideas to examine the processes of decision and deliberation associated with some accidents and disasters selected from the period 1965–75. Five examples, drawn from public inquiry reports which we have not previously discussed, are set out briefly in Table 9.1. These five examples were chosen for illustrative purposes from the sets of reports available within the given period, but they display no particularly unusual features in relation to the analysis to be presented, and other examples could readily have been chosen. The five cases all offer illustrations of unexpected or accidental events encountered by individuals within organizations in the marine, mining or construction industries, and, in the case of the Ronan Point explosion, the events also impinged upon local government officials and, of course, upon the inhabitants of the block of flats affected, this group having no relevant organizational affiliations.

Assuming that the organizations concerned were all trying to behave rationally, and that they regarded the avoidance of breakdowns, wrecks, explosions and so on as aims to be pursued rationally, it is clear that the forecasting of the future was inadequate in each case, since none of the incidents was foreseen. To this extent, the intention of those trying to behave rationally was not fulfilled, and we may ask how this failure occurred in each case. We can also ask how the organization or organizations concerned responded to the new information supplied by the news of the accident.

We shall attempt, therefore, to concentrate upon the intendedly rational features of the events which preceded the five incidents, setting on one side the mass of qualifications which must be made about the

TABLE 9.1. *Five instances of unexpected events.*

I. *Anzio I* [17]

At 1450 hours on 2 April 1966, motor vessel *Anzio I* encountered a Force 6 gale, which by 1800 hours had increased to Force 8. At 2325 hours, the vessel was observed by coastguards, and seen to be in difficulties. The vessel was beached, and vain attempts were made to fire a line to the ship. The prevailing conditions prevented any rescue attempts being made, and the master and the crew of 12 died.

II. *Starcrown* [18]

On Christmas Day 1963, the Master of motor vessel *Starcrown*, en route from Nova Scotia with a cargo of grain, took over the wheel from his third officer. Although the Master, a man with a previous record of good seamanship, was unsure of his position relative to the nearby coastline, he did not check his position, but 'inexplicably' took a dangerous course, and ran the vessel aground. No lives were lost, but the vessel was a 'constructive total loss'.

III. *Isle of Gigha* [19]

On 11 November 1966, motor vessel *Isle of Gigha*, a purpose-built ferry carrying two lorries in the vicinity of the Inner Isles of Scotland, experienced rough weather, the lorries and their cargoes moved and the ferry overturned, with the loss of two lives.

IV. *Markham Colliery Overwind* [20]

Eighteen men were killed and eleven seriously injured in a pit cage accident at Markham Colliery, Derbyshire on 30 July 1973, when a component in the winding gear failed and the cage fell back down the shaft.

V. *Ronan Point* [21]

Ronan Point is a now well known, 64-metre high block of flats in East Ham, London. On 16 May 1968 a leakage of gas from a defective component in Flat 90, on the 18th floor, led to a gas explosion, and the subsequent progressive collapse of part of the building caused the death of four residents and injured a number of others.

---

irrational elements present in all human behaviour, and about the tendencies displayed by all individuals to rationalize both their past experiences and their future plans. We may then ask what lessons the five cases have to offer us about the nature of perceived departures from intended rationality.

9.5. *Perceived departures from rationality*

We can use these cases as a basis for a discussion of departures from rationality if we make the prior assumption that one of the concerns of courts and tribunals of inquiry is to assess, with the benefit of hindsight, the nature of the relevant decisions preceding the incident being investigated. We may then take this assessment as an indication of the points at which departures from rationality had occurred, and at which failures of intended rationality had been experienced by the people concerned [22].

Thus, reviewing the five cases set out in Table 9.1, and considering the conclusions offered by the relevant inquiries, we find that the loss of *Anzio I* was attributed by the Court of Inquiry principally to the advent of exceptionally severe weather conditions which overwhelmed a seaworthy and competently-run ship, and no recommendations for changes in practice were made as a result of this inquiry. The inquiry into the grounding of m.v. *Starcrown* found that the Master's navigation was on this occasion 'deplorable', and he fully admitted blame. The overturning of the ferry *Isle of Gigha* was primarily due to a failure to secure the lorries on board, combined with a number of other small factors, but the Court also found that the owner's praiseworthy idea of having a ferry specially built for the Inner Isles, off the Scottish coast, ran into difficulties because the designers of the ship had insufficient knowledge to design a vessel which would be small enough to navigate the Crinan canal, while being seaworthy enough to operate in the open seas off Scotland.

In the remaining two cases, the respective inquiry reports set out much more extensive series of recommendations as a result of the investigations carried out. After the Markham Colliery incident, when a component in the winding gear of a pit cage failed, recommendations were made that mandatory engineering instructions should be enforced more stringently, that similar winding gear at other pits should be modified, that maintenance procedures should be reviewed, that the training of winding engineers should be improved, and that steps should be taken to eliminate or to make 'fail-safe' any single line components in winding gear. In the series of recommendations from the Ronan Point inquiry, it was suggested that the likely effects of future explosions and other possible impacts upon blocks of flats should be reduced by viewing tower blocks as civil engineering structures (which had not been done before); that the design of system-built blocks of flats should be improved; that existing blocks of flats should be strengthened; and that the Building Regulations should be modified to deal with new modes of construction.

Making the assumptions, for our present purposes, that the public inquiries concerned were able to satisfy themselves reasonably as to the events leading up to the accident in question, and that they were not biased in any major way by a desire to 'whitewash' the incident or to protect vested interests [23], we may take the findings of the inquiries as offering some indication of the manner in which the intention to pursue fully rational behaviour was inadequately realized. Demonstrating that this assumption did not hold in a particular case would not, however, invalidate the analysis to a great degree, for even if an inquiry had been prejudiced and rigged, it is still reasonable to assume that in a society which values rationality highly, the members of the court or tribunal would have felt a need to present their findings or conclusions, blinkered though they may have been, in a 'rational' manner, in accordance with an accepted model of how organizations and the individuals within them behave, or might be expected to behave. Whether these assumptions hold in each particular case or not, the person reading the inquiry report must decide for himself for each incident if the rationality framed in the inquiry conclusions is closer to 'global' rationality than was that of the organization or organizations originally concerned.

We have in Table 9.1, then, five incidents which may be treated *prima facie* as instances of failures of rationality. But merely attaching this label to them does not help us very much to distinguish between the different kinds of failure displayed in these cases, and it is in this aspect that the interest of the five examples lies. In order to develop our understanding of the different ways in which these various organizations failed to deal adequately with their environment, we need to make use of our discussion earlier in the chapter in which we presented organizations as hierarchies of decision-premises, mobilized in order to cope in an intendedly rational manner with events generating threatening amounts of uncertainty.

Using this analysis, we find that it is possible to distinguish between the five incidents in Table 9.1 by classifying them according to the change in decision-premises induced in each case by the unexpected event. The results of such a classification are summarized in Table 9.2.

Looking at the first incident, the loss of *Anzio I*, it may be recalled that the inquiry found no factors which were felt to lead to a revision of existing sets of decision-premises. The ship was seaworthy, the captain and crew competent and nothing ascertainable was wrong, so that no improvements in intendedly rational behaviour could be offered which would have avoided the incident, or which would avoid a similar incident in similar circumstances in the future [24]. In the second incident, the loss of *Starcrown*, the Master had his navigational decisions criticized as 'deplorable', and accepted the blame. Thus, here, although the individual acknowledged a depature from 'intendedly rational' behaviour, no revision of the officially accepted and institutionalized decision-premises

DISASTERS AND RATIONALITY IN ORGANIZATIONS 175

TABLE 9.2. *Table of unexpected events discussed.*

Incident	Inquiry findings	Norms for future 'intendedly rational' behaviour derived from the incident
1. *Anzio I*	Loss due to exceptionally severe weather conditions. No recommendations made.	No improvements in intendedly rational behaviour available; no revision of decision premises possible.
2. *Starcrown*	Master's navigation 'deplorable'—blame accepted by Master.	Revision of personal decision-premises implied in this instance, but no change in officially approved proceedures. REVISION AT ONE LEVEL
3. *Isle of Gigha*	Due to failure to secure lorries, and other minor factors, but the major contributing factor was inadequate design of the vessel.	Some revision of premises for the Master indicated, but a major revision at a second, design level also. Third level, commissioning of the design, was approved. REVISION AT TWO LEVELS
4. *Markham Colliery*	Due to failure of single line component in winding gear which was difficult to inspect and maintain.	Revision in five decision zones indicated: —enforcing mandatory engineering instructions; —modifying maintenance procedures; —improved training for winding engineers; —different action by winding engineers; —change in design criteria. *No change* for miners REVISION IN FIVE ZONES

TABLE 9.2.—(Cont'd)

5. Ronan Point	Gas leakage due to defective component caused explosion which led to progressive collapse of corner of block of flats.	Revision recommended at several levels of decision zones, including all of those concerned with engineering aspects of high, system-built buildings: —gas authorities and fitters; —architects at many levels; —construction engineers; —local authorities at many levels; —civil servants responsible for framing and updating Building Regulations. *No change* implied for decision-premises of victims or other residents. REVISION AT MANY LEVELS

about how to handle a merchant ship was felt to be called for.

These first two cases offer instances in which there was 'no change in decision-premises', and 'a change only in the decision-premises used unofficially by one individual, at one level'. The remaining cases all involve more than one level in the decision hierarchy, and also involve some shift in officially approved bases for decision-making. The *Isle of Gigha* inquiry deals with decisions at two levels, approving the decision-premises at a third level. The Master was under some criticism for failing to secure the lorries on the ferry, but the main recommendation for the revision of decision-premises was directed at the designers who had inadequate knowledge of the operating conditions for which their vessel was designed. The owner's decision to commission the building of a ferry for the Inner Isles was endorsed by the Court of Inquiry.

The fourth case presented in Tables 9.1 and 9.2, the accident resulting from the Markham Colliery overwind, produced recommendations for the revision of the existing decision-premises used by those responsible for enforcing certain mandatory instructions; by those responsible for other similar winding engines, and for their maintenance; by those responsible for training winding engineers, and, by implication by winding engineers themselves; and finally revision of the decision-premises used by the designers of winding equipment. As a result of the first inquiry report into this accident, then, revisions were recommended of five distinct, though linked decision zones [25].

We might note in passing that we are taking these recommendations for change, or lists of 'lessons to be learned' produced by public inquiries as indicators in themselves of the manner in which the organizational structure permitting the accident to occur is viewed by the inquiry, when assessed 'according to the norms of rationality'. Thus the discussion does not depend upon whether these recommendations are necessarily implemented, for the question of whether such recommendations are, in fact, incorporated into the collective consciousness of decision-makers is a separate question. Many accidents do, of course, arise in situations where desirable courses of action have already been suggested by earlier inquiries but have not been implemented.

The fifth example, the Ronan Point collapse, conforms to the earlier pattern of the *Isle of Gigha* and Markham Colliery incidents in that the inquiry report offers recommendations for the revision of decision-premises at several levels. Thus it was suggested that amendments be made in the assumptions underlying decisions taken by gas fitters and gas authorities; by architects designing such buildings, and by construction engineers; by local authorities; and by those responsible for the framing and updating of the Building Regulations. But in addition to this familiar pattern, the complexities of the Ronan Point case draw our attention to some features which we have not encountered in the earlier cases.

One of these features concerns the nature of the response of innocent victims to such incidents. The victims of the Ronan Point accident, like individuals involved in many accidents, went through a distressing experience in which their everyday assumptions about the way in which the world might normally be expected to behave were shattered in a major and completely unexpected manner. But, because those affected played no part in bringing about the accident, they could draw no effective conclusions from their experiences which would enable them to change their behaviour to avoid a repetition of such an incident. If they wished to behave, we might say, in accordance with an intendedly rational model of behaviour in the terms discussed above, they should continue their way of life unmodified. As a result of the inquiry, for example, it is not suggested to those people injured whilst in their living rooms at Ronan Point that they should no longer enter their living rooms, or even that they should no longer live in blocks of high-rise flats, although those who decided that they did want to move would be likely to find their decision treated with sympathy and understanding.

Some victims of tragic accidents feel very strongly that they *should* reorder their actions and their way of life after the accident, even though they and others may recognize that there is no external rational need for such a change. For example, one survivor of a particularly horrific nightclub fire in America, who had climbed out over the bodies of other victims, was reported twenty years later to be unable to tolerate carpets or soft furnishings in the apartment to which she confined herself [26]. She, and the inhabitants of Ronan Point, were involved in accidents as private individuals and as victims, but if such a strong urge to change behaviour patterns were observed in an individual who occupied an official position in an organization, in officially relevant contexts after an accident, his behaviour would be regarded not merely as 'irrational though understandable'. In addition, moves would probably be made within the organization to prevent the interaction of his feelings with his official role from interfering with the pursuit of his duties, so that, for example, he might well be removed from office on the grounds that he had 'lost his nerve'.

The second feature of general interest displayed by the Ronan Point case is the interorganizational nature of its origins. The incident did not spring from erroneous assumptions made by a single individual or work group, nor even from those of a single organization. Instead, there was, as the inquiry phrased it, a 'blind spot' in the manner in which the construction of large system-built blocks of flats was viewed. All of those concerned with Ronan Point, and with many other similar blocks, failed actively or officially to consider the possibility of progressive collapse of such buildings, whether they were architects, systems contractors, local-authority officials, government employees or engineers. An indication of the extent of the coincidence of the omission of this factor from the decision-premises and from the associated world-view of so many groups of people is given by the comment in the report of the public inquiry that no English publication 'has drawn attention to the need to think of tall system buildings as civil engineering structures requiring alternate paths to support the load in the event of the failure of a load-bearing member' [27].

As a result of this catastrophe, therefore, as with the adjustment in the official perception of spoil tips which was necessary after Aberfan, there was a correction, not of a single set of decision-premises, but of a major set of false assumptions. This massive institutional bias was shared by a cluster of organizations in a particular industry, each organization being reassured, presumably, of the adequacy of its understanding of the problem by the fact that all of the other interested parties shared this mistaken perception. Therefore the inquiry felt it necessary not only to draw attention to inadequacies in individual sets of decision-premises, but also to adjust this major misperception which was spread through a complete cluster of organizations, just as the misperception of tips prior to Aberfan had permeated the coal industry.

9.6. *The structured nature of unintended consequences: anti-tasks*

Our consideration of the above five examples, then, has enabled us to think of disasters, accidents and other unintended consequences in terms of the number of errors, misconceptions and departures from subsequently defined rationality which provoke them, and also in terms of the level in a hierarchy of decision-premises at which these errors occurred. But there are further implications of this way of considering the preconditions of unintended consequences which are worth examining.

Organizational hierarchies and the hierarchies of decision-premises associated with them are not random phenomena. In spite of the ubiquitous pressures towards randomness, as we indicated earlier in this chapter, some central ordering principle is essential to any organization if we are to continue to regard it as such. But if there is an element of order associated with any organization, or cluster of organizations, this order has implications for the incidence of unintended consequences. Even if we make the starting assumption that the incidence of errors in perceptions, information-handling and decision-making will be randomly distributed amongst the office-holders in an organization, it is clear that the consequences of these errors for the organization and its environment will *not* be randomly distributed. Depending upon the point in the organization's 'planful hierarchy' [28] at which a slip or error occurs, the consequences are likely to differ.

In general terms, slips and errors which occur in the lower reaches of an organization are likely to be more modest in their consequences, while those originating at the higher levels will be likely not only to be more far-reaching in their consequences, but also to produce more complex forms of unintended consequences, for the higher the level at which an error originates, the greater chance it has of being compounded with and extended by other errors which it encounters in the course of its transmission down the hierarchy [29]. As a result, we could regard the analysis of the incidents presented above as a kind of 'unfolding' or 'decomposing' of the events leading to these major failures of foresight into numbers of constituent errors originating at different levels in the organizational hierarchies of intention. Public inquiries may thus be thought of, in one sense, as undertaking the task of reorganizing the information available to them in order to illustrate the non-random effects of errors occurring at different points in the series of decision-making levels which they examine.

Extending this notion, it is possible, in fact, to think of unintended consequences in relation to a more extended hierarchy of individual and social purposive behaviour from which they spring. In this view, the origins of unintended consequences can be located at the physiological level where physical slips and simple typing and speech errors originate, at the psychological or social psychological level from which many individual accidents attributed to 'human error' spring, up to organizational processes which may be corrupted at many stages in the initiation or execution of organizational actions, to lead to disaster [30].

The central point here is that the more extensive a negentropic order-seeking system becomes, the greater is the potential which it also develops for the orderly dissemination of unintended consequences. Transferring to this context a principle which was formulated in the study of speech errors [31], we may put forward the general principle that 'unintended consequences produced within organizational settings make non-random use of the rules of organization in their propagation'. Thus, we are drawing attention to the fact that any organized, negentropic mode of relating to the environment creates the possibility that the organizing element may operate to produce or magnify unintended consequences in a surprisingly ordered way [32]. All that is required is the introduction of unintended or unforeseen variety near to the organizing centre to produce a large-scale, but orderly error which makes use of the amplifying power of any ordered organizational hierarchy. If we consider organizational hierarchies as systems set up to carry out tasks, these ordered but undesired consequences could be regarded as 'anti-tasks' rather than as completely random errors.

Large-scale disasters need time, resources and organization if they are to occur—if the 'anti-task' is to be 'successfully' executed. Since these conditions are most unlikely to be met solely as a result of a concatenation of random events, we can almost suggest that simple accidents can be readily arranged, but that disasters require much more organizing ability [33]!

The contaminated fluids case discussed in Chapter 6 provides us with a very simple and clear example of an anti-task, or orderly error. The job of the distribution network being operated here was to convey sterile fluids and other materials from the point of manufacture to the point of use in various hospitals. Given an error or fault in manufacture, the anti-task which the public inquiry then had to examine was the inadvertent use of this network to carry *contaminated* fluids to hospitals. Accidents which did not make use of this organizing property, events such as, say, the inadvertent distribution of the contaminated fluids to children in schools as a sweetened drink would be more disorderly errors, involving more chains of improbable or disorderly events, and would thus be less likely to occur [34]. As another physical illustration of the point being made, it might be useful to mention a recent accident in Germany in which a train carrying petrol caught fire. The burning petrol was reported to have found its way into the nearby sewage system, so that the fire was not confined to the immediate vicinity of the burning train but, once given access to the orderly system of passageways established to carry sewage, could spread much further though in an orderly and to some degree predictable fashion [35]. The establishment of an orderly system to cope with the task of disposing of sewage created, at the same time, an additional cluster of orderly properties which were very well suited to the execution of the anti-task of distributing burning petrol.

9.7. *The development of disasters*

Using these rather extended discussions and considerations, we can now return to our original concern, to gain a greater understanding of the kinds of conditions which permit and foster the development of disasters and large-scale accidents. All accidents and disasters have as one of their basic properties that they are produced by or associated with a failure of intent. This property which accidents have, almost by definition, makes it possible for surprise or an overthrowing of expectations to accompany an accident, as the understanding of the world as-it-was-intended-to-be is replaced by the understanding of the world as-it-now-is.

Accidents are produced, then, by failures of intent, by errors. But the argument developed in the preceding section suggested that most errors lead only to small-scale slips or accidents and that, except in rare cases, errors are only likely to develop into large-scale accidents or disasters if they occur in an organizational decision hierarchy or power hierarchy at a point at which they are likely to be magnified, and possibly to be compounded with other smaller errors, by the operation of the normal administrative processes. They are most likely to produce large-scale accidents, that is to say, if they are linked with the negentropic tendencies of an organization or of a cluster of organizations, so as to become major 'anti-tasks' which the organization then inadvertently executes.

If a disaster is to be produced by other means, we must assume a series of random mishaps or errors which coincidentally combine to produce the disastrous effect; and, except for the case of wholly unforeseen natural disasters which arise as a result of the random movements of very large-scale natural forces, it is difficult to imagine such a concatenation of events occurring without some access to the facilities provided by human organizing propensities. The match carelessly thrown away will not burn ten-thousand wooden homes if they have not already been built and clustered together. The plotting error of the ship's navigator can only produce a shipwreck if time, energy and resources have been devoted to building the liner concerned, filling it with people and pushing it out to sea.

As we suggested earlier, the failure of intent which produces a disaster must essentially remain covert until it is too late for its effects to be averted. If the error is spotted early on, made overt and dealt with, the disaster does not occur, and if anyone is sufficiently aware of the possibilities to talk of the incident in this way, we have a 'near-miss' instead. It is therefore essential for the occurrence of a disaster that the events which lead up to it should be hidden, or, if they are visible, that their full import should not be understood.

However, it is not sufficient to suggest, as we have been doing up to this point, that misunderstandings alone can produce disasters. Misunderstandings are states of mind, while disasters are gross, unwanted deformations of the world. The missing element which links these two conditions is energy. Purposive behaviour can only be pursued if those with an intent have access to some kind of energy source which enables them to achieve the desired transformations of the world. We can only walk to the other side of town if we can cause our legs to propel us there; we can only dig a ditch if we are able to direct our energy sufficiently precisely to achieve this objective; and even purposive communication, through speech or the written word, depends upon the precise mobilization of the appropriate muscular energy. It will be worth our while, therefore, to spend some time looking at the nature of the interrelationship between energy and information, as the next stage in our search for a better understanding of the processes which give rise to disasters.

9.8. *Energy and information*

In our examination, earlier in this chapter, of five accident reports, we found ourselves concerned with the manner in which unexpected events caused new and surprising information to impinge upon established hierarchical sets of procedures for anticipating and processing certain kinds of information. An examination of the resulting conflict of information is important and not merely an abstract exercise; because, in each of these cases, the information signalled a release of some form of energy—whether in the form of kinetic energy from the waves and the wind causing the grounding or the overturning of vessels at sea; in the form of the potential energy in a cage full of men which a broken winding-gear component released; or in the form of the chemical energy locked into a gas–air mixture which, when ignited at Ronan Point, served inadvertently to release further potential energy stored in the structure of the block of flats.

In fact, we may consider all events, not merely accidental or unexpected ones, as having both an energy and an information component. Men move towards their goals by directing energy in an orderly fashion, guided by appropriate information, in order that their desired outcome is produced. Disasters, accidents and all forms of unintended consequences may be considered as the results of energy which at some point was combined with misleading or inadequate information from the point of view of the individual or the organization concerned.

Although men talk about 'creating' energy, they never create it in any real sense, for they merely take potential energy which is locked up in one form or another, whether as food or as fuel, and redirect it for their own purposes [36]. They burn coal or oil or gas, they store water or the heat from the sun's rays, or they 'harness' the energy found within the atom [37].

When accidents and disasters occur, therefore, either energy which was to be released without human intervention anyway is released in a form which causes death, injury or loss of property to human beings, as when an earthquake occurs, or a volcano erupts; or, energy which man was harnessing for his own purposes 'slips the leash' because of some feature which man *did not know about*, and produces its own crop of disastrous events. Thus disasters are occasioned by energy plus misinformation.

The energy component of accidents and disasters has not gone unnoticed in the past, and indeed, an American accident specialist, Haddon, has drawn particular attention to it, seeking to use an analysis of the energy source giving rise to an accident to classify both the kinds of accident which occur and the range of strategies available for avoiding these accidents or for ameliorating their consequences [38]. Thus, in the study of injuries sustained by old people in falls, he identifies the source of energy as the kinetic energy of the person's body hitting an obstacle, derived from the potential energy which the body possessed before the fall. Constructing an aetiologic classification of the sources of such types of accident, he is then able to consider in a systematic manner the various options for prevention: such as whether those at risk need to stand up; whether they have to be lifted in situations where there is a danger of falling; whether the objects that the victim is likely to hit could be padded to absorb the energy of impact; whether the person likely to fall could be padded; and so on [39]. Haddon has developed this concern with the energy component further, but the preceding analysis would suggest that he has been considering only one half of the disaster 'equation'—he has neglected to link his discussion of energy with a discussion of the information or misinformation associated with the direction of that energy. Energy and information are the two elements needed to bring about desired purposive transformations of the world; and, in consequence, they are also the two essential elements needed to describe the way in which inadvertent, unintended transformations of the world come about [40].

9.9. *Transformations of energy and information systems*

Disasters arise when there is some discrepancy in this process of mobilizing energy to transform the world, a discrepancy between where the energy was intended to be directed, and where it is actually directed. The impact which an error has upon the world arises at one of the points where there is such a discrepancy. If we think of those with purposive intent as operating with some kind of map of the world which guides their actions, an accident will occur at one of the points at which their 'map' is defective. To illustrate this point by reference to actual maps: we can see that the misunderstanding which is inherent in a faulty maritime chart which fails to indicate to the master of a ship the presence of a dangerous shoal will produce disaster if the master links this faulty information with the energy which he has made available to him to propel his large, purpose-built ship through the water; and, instead, inadvertently uses it for the execution of the anti-task of ramming the ship against the submerged rocks.

Accidents are produced as a result of the combination of misinformation or misunderstanding with sufficient energy to produce an undesired transformation. Disasters are accidents which are more surprising or more alarming than usual, and, as we have seen, this depends very much upon a number of subjective elements which are inherent in the term 'usual'. Making allowance for these subjective elements, however, we can readily understand, in the light of the above discussion, why it is that organizations come to be so frequently, and indeed unavoidably, involved in any discussion of disaster. This is because organizations have two relevant properties: they are able to take misunderstandings or misinformation and amplify their effects in a way which is far beyond the capabilities of any single individual; and they are likely to be able to combine these misunderstandings with an access to transformative sources of energy which are not likely to be available to the isolated individual. Just as the control of manpower, materials and energy over long periods of time gives organizations the ability to execute tasks of a scale beyond the reach of the individual, so the same properties enable them to carry out anti-tasks of a larger scale.

As a result of these combinations of misinformation and energy, then, the world comes to be changed in unanticipated ways. But, as we have seen at a number of points in earlier chapters, it is not merely the fact that the world is transformed that we need to note when a disaster arises: because the events concerned are unexpected, there is also a transformation of that understanding of the world held by those who were expecting something else. Anyone who is surprised by an accident or a disaster needs to achieve a new understanding of the world which accommodates their surprise.

We may use this point to look briefly at the kinds of system which we are examining when we are studying disasters, and also, incidentally, to gain an understanding of the reasons why it is appropriate to use Thom's mathematical theories, which were developed to explain transformations of organic, purposive entities, for the study of disasters, which are, on the face of it, non-organic.

We have progressed, in our examination of disasters, from a consideration of disruptive external physical events to discussions of the reactions of individuals and of organizations to these events, and to discussions of novel and slightly odd concepts such as 'anti-tasks'. Thus, beginning with physical disruption, we have come to find ourselves concerned with expectations about events, with the overturning of these expectations, and with surprise and disclosure.

That is to say, in seeking for a theory of disaster, we have found ourselves concerned, not merely with a system of physical events—

SYSTEM A	PHYSICAL EVENTS, WHICH MAY CONTAIN SOME ORGANIZED ANTI-TASK ELEMENTS

Instead, we are concerned with a larger system which includes not only physical events, but also the perception of these events by individuals. Thus, to understand accidents, we need to look, not at the rather limited systems of the type represented by System A, above, but at more extensive systems containing the elements shown in System B:

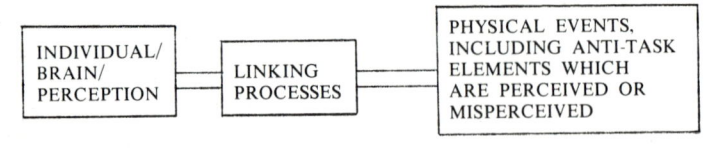

```
┌─────────────┐   ┌──────────┐   ┌──────────────────────┐
│ INDIVIDUAL/ │   │ LINKING  │   │ PHYSICAL EVENTS,     │
│ BRAIN/      │───│ PROCESSES│───│ INCLUDING ANTI-TASK  │
│ PERCEPTION  │   │          │   │ ELEMENTS WHICH       │
└─────────────┘   └──────────┘   │ ARE PERCEIVED OR     │
                                 │ MISPERCEIVED         │
                                 └──────────────────────┘
```

SYSTEM B

And in fact since we will, in most cases of disaster, want to look also at organizational involvement and at organizational surprise, we will want to think of disasters in terms of further extended systems of the type shown in System C:

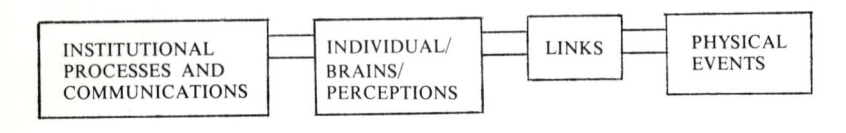

```
┌──────────────┐   ┌─────────────┐   ┌───────┐   ┌──────────┐
│ INSTITUTIONAL│   │ INDIVIDUAL/ │   │ LINKS │   │ PHYSICAL │
│ PROCESSES AND│───│ BRAINS/     │───│       │───│ EVENTS   │
│ COMMUNICATIONS│  │ PERCEPTIONS │   │       │   │          │
└──────────────┘   └─────────────┘   └───────┘   └──────────┘
```

SYSTEM C

Setting out the systems which we are studying in the above manner may help to explain why we were able to suggest that it might be appropriate to apply Thom's catastrophe theory to the kinds of transformation with which we are concerned. If a transformation is merely a random event, then it seems unlikely that it will be possible to produce a mathematical theory which will be able to describe and predict its properties, except in stochastic terms. Thom's theories imply some essential central integrity of the organized properties of the system in question. He produced his theories with the needs of biologists in mind; and, in the case of organic transformations or 'catastrophes' of the kind which occur when, say, a cell divides, the organic, biological character of the object of study itself clearly provides the central integrity of the system, which persists after the transformation, and makes it possible to carry out discussions of the links between the properties of the system before and after the transformation.

In System A, above, there is no such organizing property, even though the entropic, physical events under consideration may still retain some ordered characteristics impressed upon them as a result of past moulding by humans or other order-seeking entities. In Systems B and C, however, there is some central organic entity which persists through the transformation which occurs when an accident or a disaster takes place, this being provided by the individuals or groups of individuals involved in the system. These systems are orderly, not merely because they contain

DISASTERS AND RATIONALITY IN ORGANIZATIONS 187

some 'anti-task' elements, but also because they contain order-seeking beings. Thus, we can regard our use of Thom's discontinuities as being concerned with 'catastrophe-like' shifts in the topology of systems such as System B or System C, at a point at which surprise or disclosure occurs.

9.10. *Summary*

We have now extended our previous discussions of the limits of decision-making and their association with order and disorder from an individual to an organizational level. In tackling the question of how organizations pursue an intendedly rational path, we found that, first of all, we had to examine the way in which organizations coped with the diversity of their individual members, and the diversity of their environment, by using culture, hierarchy and power to retain organizational integration. The individual decision-maker, and the bounded rationality which his finite character imposes upon him, was tied in to the larger but still limited rationality of his organization by his location within a bounded decision-zone, where he and his colleagues tended to share views of the world, and of the factors to be considered important in decision-making.

We then examined five instances of accidents which had been the subject of inquiries, and suggested that the recommendations from the inquiries could be regarded as 'decompositions' of the points at which the accident had developed in the decision-hierarchy of the organizations concerned. Accidents and disasters thus vary in the complexity of the combinations of errors at different levels in the decision-hierarchies which produced them. Also, those errors which arise at higher levels within an organization are likely both to be more far-reaching, and to be associated with more complex kinds of accident. They are likely to be more far-reaching because the higher an error is when it occurs, the more likely it is to be disseminated through the amplifying power of the organization; they are likely to be more complex because higher-level errors are more likely to pick up and combine with smaller, lower-level errors that, by themselves, would not have produced anything very untoward.

We concluded by suggesting that accidents and disasters arising in this way may be regarded as a result of the combination of energy and misinformation, with organizations being particularly important in providing access to the energy concerned, and in supplying or magnifying the appropriate pieces of information. If we take a look beyond the material transformations associated with a disaster, it becomes clear that disasters may be regarded as transformations of socio-technical systems, in which the perceptions, expectations and understandings of those associated with the material world concerned are an integral part of the phenomena under study.

Having taken this information-based view of disasters, we have been able to bring a degree of conceptual clarity to a notoriously vague area of analysis, by relating the informational preconditions and consequences of disasters to the kinds of system-transformation discussed. By using this understanding, it should be possible in the future to look more systematically and in more detail at the social circumstances which produce, foster or inhibit the kinds of informational preconditions which create disasters.

10 The origins of disaster

We set out at the beginning of this book to try to gain an understanding of disasters and of the pre-conditions which permit their development. This aim led us to examine a range of writings on accidents and disasters, to analyse the findings of numerous tribunals and courts of inquiry, and to examine the kinds of transformation which are particularly associated with the moment when it is realized that an unexpected, an unanticipated disruptive event is about to occur or has just occurred. In the course of our examinations, we have developed and made use of a whole range of somewhat novel theoretical ideas which have helped us to begin to challenge the long-held view that all disasters are unique, by offering a set of ideas and categories which may be used to order information relating to accidents and disasters, and indeed, to a wide variety of unintended consequences.

These newly formed theoretical ideas and categories are to be found throughout earlier chapters of this book, at the points at which they have been developed. It is intended here to offer the main ideas in a shorthand form which will conveniently summarize for the reader the major points which have emerged from the research [1].

In order to understand and begin to analyse the ways in which the world is inadvertently transformed when accidents and disaster occur, we have suggested that we must pay attention to the two elements which are basic to all transformations of the world: energy and information. All disasters may be regarded as the outcomes of misplaced or misdirected energy, so that we may state the general principle that:

disaster equals energy plus misinformation

If this proposition is accepted, it becomes clear that in order to understand the origins of disaster, it is necessary to study both the social and the non-social sources of both the energy and the misinformation which combine to produce disasters.

10.1. *Energy and misinformation as sources of disaster*

Considering first of all the energy component in the above 'equation', it is clear that disasters are produced only as a result of the discharge of energy. If the energy is not discharged, the disaster does not occur. When disasters take place, therefore, one of two things happens. Either energy which is released naturally is discharged when men or their belongings are in a position to be damaged by it; or energy which man was harnessing for his own purposes 'slips its leash'.

Disaster needs energy, then, and one form of classification of disastrous events is by means of the source of energy released. By this means, it is possible to distinguish in a preliminary manner between natural disasters:

—which arise from the discharge of energy from geophysical sources;
—those such as landslides and avalanches which arise from the interaction of the earth and the atmosphere;
—those deriving from meteorological and hydrological sources of energy;
—and those deriving from biological sources, such as locust swarms and epidemics.

Man-made disasters may then be regarded as arising from:

—the energy deployed for military purposes in war;
—from impacts;
—from the collapse of structures;
—from explosions and fires;
—or from chemical or biological sources [2].

The usefulness of such a classification is limited, however, because the multiple forms of energy commonly released in many accidents and disasters complicate the pattern, and in any case, the distinction between 'natural' and 'man-made' disasters is one which is becoming increasingly blurred, as man intervenes more in his environment. But the major limitation of such classifications of disasters is that they are based upon only one of the two elements in our equation, and a more useful understanding can only be achieved if a consideration of the energy source is combined with some assessment of the informational element.

Disasters are misplaced energy, unwanted energy, just as misplaced music is noise, misplaced flowers are weeds and unwanted order may masquerade as randomness. And for this reason, we need to become aware that discussions of purpose, and of order and disorder, negentropy and entropy are impossible to avoid in discussions of disasters. Hidden in the equation set out above are a number of implied questions, for we need to know for whom the energy is misplaced, just as we need to know for whom a given set of sounds is noise, and for whom it is music. There may be disagreement among different groups about whether a particular plant is a weed, and what may look like a random number to one individual will be recognizable to another as his telephone number. For this reason, disasters can only be understood in terms of the relationship between an energy discharge and the purposes and intentions of living 'negentropic' beings. In defining a disaster, we cannot escape the need to identify the social sources of the point of view which declares the energy to be misplaced.

From the point of view of the victim, of course, the energy is always misplaced, and for many purposes the victim will be indifferent to the sources, when his concern for the effects is used as a standard of comparison; thus he may be under strong pressure to class together as disasters those events which cause damage and distress for him, whether these flow from natural events, from the mistakes of the well-intentioned, or from the successful execution of the intentional plans of a terrorist or enemy.

For those who deal with post-disaster plans, too, and for those who coordinate emergency services and treat the injured, the question of the source of the destructive energy is secondary, for the same kind of energy discharge produces the same kind of damage and injury, regardless of the will, good-will or lack of will behind it.

But if we look at this question from the point of view of the knower, the user of the information or misinformation, we find that we have no problem in explaining the origins of deliberate acts intended to provoke the release of energy to the damage of others. The problematic cases are those where the energy release is unintentional, whether the failure of intention lies in not maintaining adequate control over energy harnessed by men; in not reading adequately and sufficiently soon such warning signs as are offered by the advent of natural discharges of energy; or in not acting on them sufficiently promptly or appropriately.

We seek to create order, as 'order-eating' beings [3], and since we seek to order our world with *our* order, we always run the risk of clashing with those who have alternative visions, so that the question of power between various individuals raises itself at a very early stage in our argument. If we had perfect information about the world, if we were omniscient, we would be in a position to achieve all of our goals, provided that they were physically possible, and since we could impose our own order, we would be all-powerful. But humans are not in such a position, for they can never be sure of typing a note, or speaking an order, or executing a policy without running into unforeseen energy discharges from a variety of sources which act to subvert the intended goal to some degree, whether these sources arise from unknown, inanimate objects, or from purposive entities trying to impose their own order (whether these latter are termites chewing up our floor, robins nesting in our car engine or saboteurs blowing up our factories).

Where our purpose or intention is not going to be fulfilled as a result of factors of which we are unaware, we will realize, at some point, the inadequacy of our information—either at an early stage, when we can take some corrective action; or at the stage when a disaster begins to occur, when it is too late, although we may still be able to ameliorate the consequences, or prevent similar occurrences in the future, if the information we have gained is of the correct kind to make this possible.

When an accident occurs because of what we could call a 'lapse from good practice', we are implying that there exists a body of knowledge about the particular hazard which is generally accepted, and that the accident in question arose because some individual or group failed to take account of, or chose not to take account of, the information available. In such cases, we may attribute the source of the accident to the fact that because of ignorance, impatience or oversight, the walker failed to check the weather forecast before setting out across the moors, say, or the captain failed to keep a look-out while navigating crowded waters. In such cases, the individual discontinuities in the understanding of the manner in which the world operates, on the part of those involved, are adjusted with very little impact upon the more widely accepted view of the world. Society, in such cases, can say, 'I told you so', and take the accident as a warning of the consequences waiting for those who ignore accepted good practice.

But when, at a collective or societal level, the accident is less expected, we are not in such a superior position, for the collective, intendedly rational assessment of the situation and its likely risks by a group of which we are a part has been at fault. Then the manner in which we absorb and process information about the world is abruptly brought into question, either because what had previously been uncertain is now suddenly made clear, or because what we had previously relied on as being clear has now become problematic. As an aside, we may note that all of those factors which sociologists and philosophers of science have discussed as being relevant to the process of discovery in science could be brought to bear at this point; for they are all potentially useful in understanding the changes which groups undergo as they strive to analyse the world in which they are operating, even though in this case the groups are likely to be made up of administrators rather than scientists.

10.2. *The incubation of disasters*

Beginning with a definition of disaster which stresses information and misinformation, we were encouraged to look, not as so many of the previous workers in this field have, at the physical disruption which flows from a disastrous event, and at the social reactions to this, but rather at the cultural disruption which is produced when anticipated patterns of information fail to materialize [4]. This drew our attention, not to the way in which groups and individuals cope with and react to misfortune, but to the manner in which they gradually come to develop and rely on a mistaken view of the world, and to the events which lead to this error being unmasked.

We began by imagining a situation where a community or group were originally possessed of sufficiently accurate information about their surroundings to enable them to construct precautionary measures which successfully warded off known dangers, to provide us with a 'notionally normal' starting point for the development of disasters. From this starting point, for each disaster or large-scale accident which emerges, we have suggested that there is an 'incubation period' before the disaster which begins when the first of the ambiguous or unnoticed events which will eventually accumulate to provoke the disaster occurs, moving the community away from the notionally normal starting point. Large-scale disasters rarely develop instantaneously, and the incubation period provides time for the resources of energy, materials and manpower which are to produce the disaster to be covertly and inadvertently assembled.

At the end of the incubation period, some unambiguous 'precipitating event' [5] which cannot be ignored makes clear the discrepancy between the environment as it had been believed to be, and the environment as it actually is, forcing a reassessment and reinterpretation of the various hidden, ambiguous or anomalous events which have accumulated during the incubation period. This reassessment may involve sudden 'gestalt shifts' in the understanding of the situation on the part of many of the individuals concerned, particularly the unsuspecting injured victims; but it is important to realize that the process which is central to this approach to the understanding of the origins of disasters is one which produces a similarly abrupt shift in the collective, institutional assessment of the hazard producing the disaster. As an indication of one avenue for future research, we may note that it would be possible to treat the information-accumulation process, and the sudden shift in information levels associated with the precipitation of the disaster, as an elementary 'catastrophe'—using this term in the mathematical sense developed by Thom [6], so that there is a possibility of developing a mathematical treatment of the information processes being described.

10.3. *The prevention of disasters*

It is, of course, rather an obvious point to make, that if the first item which constitutes the start of an incubation period were noticed, and its implications fully understood, and if all subsequent items were dealt with in this way, the disaster would never happen. To achieve this outcome, as we have already commented, we would have to be omniscient, so that it is of little use to approach the prevention of disasters in this way. What the analysis presented does enable us to do, however, is to examine the range of events and errors which would have to be coped with if disasters were to be prevented [7].

To prevent accidents and disasters, we would need to be aware of each point in a developing incubation network so that we could clarify ambiguities, make sure that information is not overlooked, and provide both the information to control known complex situations where energy discharges might occur, and the information needed to foresee unknown, unprovided for, situations. We would need continuously to adjust the incipient discrepancies between the picture of the world envisaged in someone's plan, and the way the world really is. We would have to provide a perfectly accurate, continuous feedback which would ensure that all informational adjustments made by those pursuing the plan would be infinitely small shifts of view, so that the information-acquisition process proceeded perfectly smoothly. When such a matching is not possible, discontinuous jumps in the acquisition of the relevant information will occur.

Many errors and accidents, of course, could only be stopped if we were able in some way to operate inside the bodies of various participants, spotting faulty nerve-cell discharges and potentially dangerous muscular twitches; or we would have to operate, for example, in close collaboration with the driver of a car who was about to become involved in an accident and advise him of the precise evasive action to be taken. Since we are unable to intervene directly in this way, we seek to achieve the same end by trying to provide reliable muscle action through appropriate training, of the kind offered to typists or to tight-rope walkers; or by providing additional information through instruments which tell a driver of the presence of ice on the road. A proposal has recently been made that the operators of chemical plant should have computer assistance in deciding what action to take in case of an unexpected breakdown, in order to avoid worsening the situation and leading to disaster; chemical plants are now so complex that the operators by themselves may have insufficient information to allow them to deal effectively with unexpected contingencies [8].

To combat many other features which arise during incubation periods, however, the action called for is less physiological or psychological than social. And, particularly where incubation periods extend over months or years, there is less need for split-second action than for good advance information: which warns the building designer that there is a fault in his constructional system; the gas fitter that he is about to install a defective component; or the G.P. that his patient has been exposed to an infectious disease.

We can therefore ask the question, which turns out to be in large part a sociological one: what stops people acquiring and using appropriate advance warning information, so that large-scale accidents and disasters are prevented? The answer to this in general terms is that the relevant information is not available to them at the appropriate time in a form which it is possible for them to use.

The relevant information needed if we are to prevent disasters may therefore be divided into the following groups:

(1) that which is completely unknown;
(2) that which is known but not fully appreciated;
(3) that which is known by someone, but is not brought together with other information at an appropriate time when its significance can be realized and its message acted upon;
(4) that which was available to be known, but which could not be appreciated because there was no place for it within prevailing modes of understanding.

Each of these factors may now be discussed more extensively [9].

10.3.1. *Completely unknown prior information*

Where the information which foretells disaster is completely unknown and unsuspected, it is clear that there is little that can be done except to note that better search procedures might be employed in the relevant area. But, in fact, disasters arising from such completely unknown, unsuspected sources seem to be very rare indeed, particularly in the twentieth century, and in most cases the prior information can be dealt with under the remaining three headings.

10.3.2. *Prior information noted but not fully appreciated*

Where information is potentially available, but not fully appreciated, we need to ask why it is not correctly perceived or fully understood. The kinds of answers to this question which emerge from the analyses carried out suggest that the information may not have been grasped completely because individuals, groups or institutions have a false sense of security or invulnerability when faced with danger signs; because pressure of work or other distractions draw their attention away from the emerging signs of danger; because they distrust the source from which the warning apparently comes; because they are 'decoyed' into concentrating upon one property of an emerging phenomenon and so neglecting other features of it; because they have difficulty in classifying the phenomenon and in deciding upon a suitable course of action; or because they have difficulty in identifying the information-providing event among a mass of irrelevant material and in separating it from the surrounding 'noise'.

10.3.3. *Prior information not correctly assembled*

In the third case, where information about an impending disaster is available in the hands, minds or files of individual humans, or of the institutions to which they belong, it appears to be an even more pertinent question to ask why the information is not brought to a place where it can be properly assessed, and responded to in an appropriate manner. There are a variety of possible responses to this question:

(*a*) *Information buried amongst other material.* It may be, in some cases, as it was with the information which the American intelligence services had about the impending Pearl Harbor attack, that the information is too dispersed, and surrounded by too much other, apparently relevant material to enable it to be confidently selected and centrally assembled in the time available. In smaller-scale situations, it may still not be possible to assemble the warning information because of a failure to realize the significance of one particular portion of it, or because of a failure to convince those in power, those in the most influential positions, of the validity of the information.

(*b*) *Information distributed among several organizations.* Very commonly, too, the available information may be shared out among different organizations and institutions so that, if it is to be brought together, use may have to be made of non-routinized and non-institutionalized patterns of communication. One case which we have considered where such communications did take place was the 1973 London smallpox outbreak; there the death toll would certainly have been much higher than two if the first phase of the outbreak had not been brought to an end by the activities of the doctor who resorted to the highly unorthodox expedient of making use of visiting hours to take scrapings for analysis from the blisters of a patient under someone else's care, in a public ward; and if the second phase of the outbreak had not also been limited by the exercise of an unusual degree of initiative by a social worker. Both of these unusual acts led to the discovery and containment of previously unsuspected cases of smallpox in public places. If they had not taken place, the London smallpox outbreak would have been much more extensive and damaging before being diagnosed, fully assessed and brought under control [10].

Where the pattern of routine and regular communication sets up boundaries and barriers to the flow of information relevant to the understanding and prediction of the emergence of hazards, these barriers may make it difficult to stop a disaster occurring. There is a dilemma, of course, about this issue, for it is not possible to ensure that everyone communicates everything to everyone else; and the effectiveness of all organizations and institutions lies, in part, in the manner in which they restrict communications to those which they believe to be the most essential.

(c) *Limited information available to two parties.* There is a class of accidents and disasters caused by the interaction of two parties, each of whom would have been perfectly safe had the other not been there. The restriction of information within this class of events relates to the difficulty which each party has in correctly divining the next move of the other. When ships, trains or aeroplanes collide, in many cases information-flows within the separate vehicles, between the crew members for example, are proceeding satisfactorily, but unsatisfactory communication between the vehicles takes place on the basis of visual or radar observation, through a third party, or directly by means of restricted signalling codes.

(d) *Prior information wilfully withheld.* In the cases just discussed, information which was potentially available for the prevention of disaster was not communicated because of the action of barriers erected for other purposes. But there is a further kind of inhibition of information-flows which needs to be noted separately. We are excluding from our discussion the kinds of damage wilfully caused by enemy forces or saboteurs, but it is also possible for individuals who may not have intended to create any kind of damage or disruption to conceal or withhold information which might have brought an incubation period to an end without damage or disruption. They may have a whole range of personal or idiosyncratic motives for withholding the information; but the outcome of their action is that the incubation period is not ended before a disastrous precipitating event occurs, as a result of conscious acts of concealment, or because of some other form of inhibition of the full and open communication process which might otherwise have occurred. Thus in the case discussed in Chapter 9 for example, of the loss of motor vessel *Nicolaw* in November 1969, the Court of Inquiry into the loss made it very clear that events which were known by some of the individuals involved could have provided forewarning of a possible disastrous outcome, and that these events were effectively concealed as a result of intentional actions on the part of some of those involved. A major factor contributing to the loss of this vessel was the fact that the Master concealed, from port authorities and certain other organizations with statutory safety duties, a number of inadequacies in the safety provisions made with regard to *Nicolaw*. Had these inadequacies been known, the ship would not have been allowed to put to sea [11].

10.3.4. *No place for information in existing categories*

In all of the cases discussed above, if we had wanted to take action to prevent the disaster, we would have had to try to improve the flow of information in order to reduce ignorance about the environment in question. If we regard the relationship between an individual (or an organization) and the environment as a channel of communication, we can characterize the individual or the organization as having a framework of understanding, or a theory about the environment. Additional information received along the communication channel about the environment serves to fill in the picture which the receiver has of the environment, lowering uncertainty by supplying additional information which reduces the areas of ignorance in the receiver's understanding. In such cases, we would try to prevent disaster by improving information flows, so as to hasten the uncertainty-reduction process.

But there is another most important category of disaster preconditions which would require a different kind of action if disaster were to be avoided. Sometimes, individuals or organizations are unaware of their areas of ignorance: they operate with theories about their environment, with pictures of the world which they inhabit which have no place for the new hazard which threatens them. In these cases, the improvement of information flows is not the essential action needed to end the incubation period before a disaster occurs. Instead, the existing theory about the world must in some way be revised so that its inadequacies can be exposed.

To mention once more the example that has recurred throughout our discussions, it is clear that the National Coal Board has always been aware of its lack of complete knowledge with regard to the hazards arising from potential roof falls in mines, and those arising from potential explosions in mines; and it is clear that the Board has been concerned to acquire additional information to reduce its uncertainty about such phenomena. However, it was not aware of its ignorance about the hazards of tips on the surface: they were not institutionally recognized as hazardous, no official precautions were taken to deal with possible dangers arising from tips, and those individuals who tried to point out the dangers had great difficulty in getting their voices heard.

Where there is a danger of a disaster arising as a result of a massive institutional preconception, the action that needs to be taken to prevent it is not to improve the flow of information along existing channels, but to point to the inadequacies of these existing channels. In these circumstances, what we need is a warning of the point at which changed circumstances make it desirable for us to revise our limited view of the environment.

10.4. *Organizations and disaster*

At any of the points within the social realm where we have suggested that it might be possible to take action in order to inhibit disasters, alternative patterns of action by individuals could change the course of events. But it is an over-simplification to think of the accumulation of circumstances which provoke or inhibit disaster as growing as a result of some kind of aggregate of individual actions. Individuals *are* involved, but not many isolated, unconstrained individuals. Because of the preponderance of organizations in the modern world, and because of the near-monopoly by organizations of access to and control of most of the major sources of energy which could be discharged to create disasters, it is likely that in virtually every case of disaster there will be an organizational involvement. And it follows from this that most of the individuals concerned in the events which constitute the incubation network are likely to be operating in institutional roles.

Because of this major organizational involvement with the forces and conditions likely to produce disasters, it becomes clear that anyone concerned to understand more about the origins of disaster will need to study the forces which affect the manner in which organizations handle communications about potentially dangerous energy sources of all kinds—whether these communications are within the organizations concerned, between organizations, or between an organization and the general public. In addition, they will need to look at the manner in which organizations build up and sustain their official understanding of the environment in which they operate, and at the processes which make it possible for such an official understanding to be grossly in error.

It is said to be a characteristic of modern organizations that they are unusually rational in the pursuit of their goals, and in their manipulation of the environment to this end. But they can never be fully rational, to the extent that this implies omniscience, and they have to settle for the much more limited aim of being 'intendedly rational', acknowledging that their rationality has bounds and limits. Organizations achieve concerted action by establishing and maintaining an agreement amongst their members that certain possibilities, issues and contingencies are important and relevant to organizational decision-making; while other possibilities, issues and contingencies may be ignored without incurring official disapproval. We have drawn attention to the way in which part of this process is developed and sustained, particularly in large organizations, by the development of aspects of the organization to resemble hierarchies of what we have called 'bounded decision zones' [12]. In this and similar ways, the rationality of an organization is bounded, and because of this limitation, the possibility arises that some of the contingencies which have been ignored or set aside may turn out to be much more hazardous than had been anticipated.

10.5. *Conclusions: disasters, information and energy*

To understand the origins of disaster, it is necessary to understand the kinds of event which may accumulate within an incubation period to give rise to a disaster. But the relationship between institutions and the world may be regarded as a continual cycle of assumption; exposure of the limits of assumption; the subsequent revision of the assumptions; and their replacement by new ones. And because of this, it is also important to understand the mechanism by which this continuing process comes about.

After a disastrous energy discharge, the need to make some kind of adjustment to this unforeseen event stimulates the flow of information, both at an individual and at an institutional level. People are concerned to know what kind of event it was, what kind of energy was discharged, what its consequences were for the material and social fabric, and how these consequences can be accommodated to the existing state of affairs. Some of these patterns of communication will be those initiated by official inquiries in the course of their attempts to establish the causes, to define the points at which suitable interventions could have prevented the incident. The inquiries are thus concerned to establish how, in their judgement, the assumptions, the decisions and the behaviour of individuals and organizations contributed to the event in question, and how technical, social and administrative arrangements need to be adjusted to prevent a recurrence of similar events.

The judgement of inquiries themselves, of course, may be faulty or incorrect, so that they may try to impose unworkable conditions upon the institutions concerned, in spite of an intent to be rational. Alternatively, the inquiry may just settle for a 'whitewash job'. Nonetheless, in Britain at least, they constitute a major social mechanism for adjusting to the revelations which disasters always provide, and for trying to accommodate the lessons of these revelations into the collective experience.

The problem of understanding the origins of disaster is the problem of understanding and accounting for harmful discharges of energy which occur in ways unanticipated by those pursuing orderly goals. The problems of explaining the events preceding and surrounding disasters may then be seen as the problems of accounting for biases and inadequacies in the habitual ways of handling information relating to impending energy discharges. We can, and we do, pursue such problems with the intention of using rational means of solving them and subduing disasters. Some of the problems may arise because of our attempts to manipulate the world by rationally guided processes, but we cannot abandon the rational linking of knowledge to action. Our only option, particularly in a world dominated by large-scale organizations and dependent upon large-scale technology, is to come to a better understanding of what we can and what we cannot achieve with our knowledge; so that on the one hand we can continue to try to improve it, whilst on the other hand we do not come to rely upon it in the blind hope that it will perform miracles for us.

However comforting the promise of an infinite tidiness offered to man by the older rationalist notion of the possibility of arranging our affairs always on the basis of the anticipation which our conscious knowledge offers us [13], we must recognize that we are in a contingent universe, in which ultimately there are limits on our ability to reduce uncertainty, to master all of the open-ended and perverse qualities of our environment, and upon our ability to prevent disaster. If we start by recognizing that instability lies at the heart of the world, then we may come to realize that the optimism and the assertion of certainty which enables life to create and spread order cannot completely overcome this instability. We may come to realize that, even when our strategies are successful, they are still dependent upon the munificence of the environment and upon the mutability of fortune [14].

Appendix

List of 84 accident and disaster reports published by British Government sources between 1 January 1965 and 31 December 1975 examined in the present study.

According to *British Government Publications*, 1965–75 (London: HMSO) 449 reports were published during this period. The reports examined exclude all reports on civil air accidents (209) all but one of the reports on railway accidents (121 out of 122) all reports of accidents to fishing vessels (28 out of 60 wreck reports) and seven miscellaneous reports not included in Parliamentary papers for the period. The Report of the Summerland Fire Commission is included, although strictly it is not a British Government publication.

For further details of the method of investigation, see B. A. Turner, 1976, *The Failure of Foresight* (Ph.D. thesis, University of Exeter).

A. *Reports on Mining Accidents and Other Miscellaneous Accidents*

Explosion at Cambrian Colliery, Glamorgan, 1965 Cmnd 2813 (London: HMSO).
Report by the Tribunal of Inquiry (Aberfan), 1966–67 HC553 (London: HMSO).
Report of the Inquiry into the Causes of the Accident to the Drilling Rig Sea Gem 1967 Cmnd 3409 (London: HMSO).
Report of the Inquiry into the Fire at Micheal Colliery, Fife, 1968 Cmnd 3657 (London: HMSO).
Report of the Public Inquiry into the Accident at Hixon Level Crossing on 6 January 1968, 1968 Cmnd 3706 (London: HMSO).
Public Inquiry into a fire at Dudgeon's Wharf, on 17 July 1969, 1970 Cmnd 4470 (London: HMSO).
Report of the Inquiry into the collapse of Flats at Ronan Point, Canning Town, 1968 Ministry of Housing and Local Government (London: HMSO).
Outburst of Coal and Firedamp at Cynheidre/Pentremawr Colliery Carmarthenshire, 6 April 1971, 1971 Cmnd 4804 (London: HMSO).
Report of the Committee of Inquiry into the Fire at Coldharbour Hospital, Sherborne on 5 July 1972 1972, Cmnd 5170 (London: HMSO).
Report of the Committee appointed to inquire into the circumstances including the production, which led to the use of contaminated infusion fluids in the Devonport section of Plymouth General Hospital, 1972 Cmnd 5035 (London: HMSO).
Lead Poisonings at the RTZ Smelter at Avonmouth, 1972 Cmnd 5042 (London: HMSO).
Report by HM Factory Inspectorate on the Collapse of Falsework for the Viaduct over the River Loddon on 24 October 1972, July 1973 HC 425 (London: HMSO).
Inrush at Lofthouse Colliery, Yorkshire, 1973 Cmnd 5419 (London: HMSO).
Extensive Fall of Roof at Seafield Colliery, Fife, 1973 Cmnd 5485 (London: HMSO).

APPENDIX 203

Accident at Markham Colliery, Derbyshire, 1974 Cmnd 5557 (London: HMSO).
Report of the Committee of Inquiry into the Smallpox Outbreak in London In March/April 1973, 1974 Cmnd 5626 (London: HMSO).
Report of the Summerland Fire Commission, (undated—1974) (Isle of Man: Government Office).
The Flixborough Disaster: Report of the Court of Inquiry, 1975 Department of Employment (London: HMSO).
Report of the Committee of Inquiry into the Fire at Fairfield Home Edwalton, Nottinghamshire, on 15 December 1974, 1975 Cmnd 6149 (London: HMSO).

B. Boiler Explosion Reports

3439	Explosion from a hot water boiler in a launderette at 9, East India Dock Road, London E.14.
3440	Explosion from a laundermat hot water system at Berkeley Services (Stoke Newington) Co. Ltd.
3441	Explosion from a Lancashire boiler at Willow Dye works, Leicester.
3442	Explosion from a Sentinel Locomotive boiler at the works of Gens Mills and Co. Ltd., Middlesborough.
3443	Explosion from the main boiler of the s.t. Arctic Adventurer.
3444	Explosion from a Vertical Multitubular boiler at the works of the Butterly and Blaby Brick Company Ltd., Ambergate, nr Derby.
3445	Explosion from a main engine stop valve aboard the steam tug 'No. 3'.
3446	Explosion from the boiler of 'TIC Dredger No. 9'.
3447	Explosion from a concentrator at the works of A. H. Taylors Batches, Hide, Skin, Fur and Wool Co., Cheltenham.
3448	Explosion from No. 4 'Kelly' filter at the Alkylation Plant, Heavy Organic Chemical Division, ICI Ltd., Billingham-on-Tees.
3449	Explosion from an Autoclave at the works of Cape Building Products Ltd., Iver Lane, Uxbridge.
3450	Explosion from a laundermat hot water system at Red and White Laundries Ltd., London E.15.
3451	Explosion from a laundermat hot water system at Mansfield Cooke and Co. Ltd., London W.9.
3452	Explosion from a Danks Interflo Treble pass Horizontal Oil fired Economic Boiler at the premises of British Transport Hotels Ltd., Willesden Laundry.
3453	Explosion from a horizontal boiler at GKN Screws and Fasteners Ltd., Smethwick on 6 September 1967.
3454	Explosion from a forced steam circulation boiler at Briarsdown 'B' Power Station, Enfield, Middx.
3455	Explosion from a Self-service Launderette hot water system at the premises of Speedwash, Leigh on Sea.
3456	Explosion from a drying cylinder at the New Waterside Paper Mills, Eccleshall, Darwen, Lancs.
3457	Explosion from A Water Tube Boiler at Messrs Marsh and Baxter Ltd., Hall Street, Brierley Hill, Staffordshire.
3458	Explosion from an oil heated mixing vessel at the premises of A. Long Products Ltd., Harbour Road, Rye, Sussex.
3459	Explosion from a Danks 2 pass Horizontal Economic Boiler at the works of Stevensons (Dyers) Ltd., Ambergate, Derbyshire.
3460	Explosion from a water tube boiler primary superheater outlet header at Wakefield B Power Station.
3461	Explosion from a hot water storage cylinder in a launderette at 197, Rosemount Place, Aberdeen.
3462	Explosion from a Cochran Auxiliary Boiler on board m.v. *Wellpark* O.N. 300198, at Workington on 28 February 1971.

3463	Explosion from a steam autoclave at Superbrix Ltd., Moneymoon Farm, Canwell, nr Sutton Coldfield, Warwickshire.
3464	Explosion from a boiler at Johnson Brothers (Division Josiah Wedgewood) Alexandra Pottery Works, Scotia Road, Tunstall, Staffordshire on 29 January 1971.
3465	Explosion from the main boiler casing of the VLCC *Esso Ulidia* (O.N. 339297) on 20 October 1970.
3466	Explosion from a steam expansion piece at the works of the Radcliffe Paper Mill Co. Ltd., Johnson Street, Radcliffe, Manchester on 9 October 1972.
3467	Explosion from a cast iron isolating steam valve on board m.v. *Seconda* on 15 March 1973.
3468	Explosion from a vertical cross-tube boiler at the Dairy, 80, Cwmbach Road, Fforest Fach, Swansea, Glamorgan on 10 September 1972.
3469	Explosion from a steam boiler at Vaux Breweries Limited, The Brewery, Sunderland, Tyne and Wear on 11 October 1974.
3470	Explosion from a boiler blowdown tank at Bermuda Cleaners, 21, Rosedene Terrace, Leyton, London E.10, 9 March 1974.
3471	Explosion from a steam line at Imperial Chemical Industries Ltd., Petrochemicals Division, North Tees Works, Port Clarence, Teeside, 15 May 1972.

All of the above are reports of inquiries held under the Boiler Explosions Acts 1882 and 1890, published by the Department of Trade and Industry, HMSO, London.

C. *Reports of Courts of Inquiry into Marine Wrecks*

Fishing vessel wrecks are covered in a separate series of reports and are not included here.

8033	m.v. *Starcrown*.
8034	m.v. *Trent Bant* and s.s. *Fogo* (Portuguese).
8035	m.v. *Ambassador*.
8036	m.v. *Guernsey Coast* and s.s. *Catcher* (Liberian).
8037	s.s. *Hindsia*.
8038	m.v. *Gannet* and m.v. *Katherine Kolkman* (German).
8039	m.v. *Bow Queen*.
8040 and 8040a	Final report of Court: m.v. *Sir Joseph Rawlinson*, *Danube VIII* and *Black Deep*.
8041	m.v. *Prince of Wales* (unregistered).
8042	m.v. *Darlwyne*.
8043	m.v. *Anzio I*.
8044	m.v. *Quesada*.
8045	m.v. *Isle of Gigha*.
8046	s.s. *British Crown*.
8047	m.v. *Ardgavel*.
8048	s.s. *Hemsley I*.
8049	m.v. *Rattray Head* and m.v. *Tillerman*.
8050	m.v. *Cressington* and m.v. *Hannes Knüppel*.
8051	R.N.L.I. *The Duchess of Kent*.
8052	m.v. *Nicolaw*.
8053	m.v. *Lairdsfield*.
8054	m.t. *Esso Ipswich* and m.t. *Efthycosta II*.
8055	s.s. *London Valour*.
8056	m.v. *Bel Hudson*.
8057	s.s. *Mactra*.
8058	s.s. *Esso Cambria*.
8059	m.v. *Redthorn* and m.v. *Efpha*.

8060 m.v. *Festivity*.
8061 m.v. *Glenshiel*.
8062 m.v. *Burtonia*.
8063 m.v. *British Fern* and m.v. *Teviot*.
8064 m.v. *East Shore*.

The 32 items listed above are all reports of Courts of Inquiry carried out under the Merchant Shipping Act, and published by the Department of Trade and Industry, HMSO, London.

Notes

Chapter 1

[1] Council for Science and Society (1977) *The Acceptability of Risks: the logic and social dynamics of fair decisions and effective controls* (London, Barry Rose, in association with the Council for Science and Society); Lowrance W. W. (1976) *Of Acceptable Risk* (Los Altos, California, William Kaufman Inc.).
[2] This term is used to describe environments by Emery, F. E. and Trist, E. L., (1965) The causal texture of organizational environments, *Human Relations*, **18**, 21–32.
[3] See the discussion of environments in Easterbrook, W. T. (1966). The entrepreneurial function in relation to technical and economic change, in *Industrialization and Society* Hoselitz, B. F. and Moore, W. E., eds. (Paris: UNESCO Mouton) pp. 57–73.
[4] The properties of high-quality intelligence are discussed by Wilensky, who also argues for further research on this topic. See Wilensky, H. L. (1967), *Organizational Intelligence* (New York: Basic Books).

Chapter 2

[1] For an extensive discussion of such reactions, see Wolfenstein, M. (1957) *Disaster: a Psychological Essay* (London: Routledge and Kegan Paul).
[2] See the comments on this point by Western. Western, K. A. (1972) 'The epidemiology of natural and man-made disasters: the present state of the art' (Academic Diploma in Tropical Public Health dissertation, London School of Hygiene and Tropical Medicine, London University) Section III B. I am indebted to Western at a number of points in this chapter, and in the following one.
[3] See Fritz, C. E. and Mathewson, J. H. (1957) *Convergence Behavior in Disasters: a Problem of Social Control* Disaster Study No. 9. (Washington, D.C.: National Academy of Science/National Research Council).
[4] On this, see, for example, the discussion in Goffmann, E. (1975) *Frame Analysis: an Essay on the Organization of Experience* (Harmondsworth, England: Penguin) pp. 28–34.
[5] See Wolfenstein *Disaster: a Psychological Study, op. cit.*
[6] From among the many disaster novels written, it is worth singling out Bagley's fictional account of the preconditions, onset and aftermath of a snow avalanche, for it seems to have been based upon a considerable amount of factual research: Bagley, D. (1975) *The Snow Tiger* (London: Collins). See also Stewart, G. (1941) *Storm* (New York: Random House); (1948) *Fire* (New York: Random House); Camus, A. A. (1948) *The Plague* (London: Hamish

Hamilton); Wylie, P. (1954) *Tomorrow!* (New York: Rinehart); Wurlitzer, R. (1972) *Quake* (New York: Dutton). Lord Lytton's (1834) *The Last Days of Pompeii* (London: Walter Scott) may be regarded as an early 'disaster novel', which quite typically stresses retribution, neglecting entirely any consideration of the origins of the misfortune recounted. *Towering Inferno* is perhaps the most notable example of the 'disaster film'.

[7] Keys, A. (1950) *The Biology of Human Starvation*, 2 vols (Minnesota: University of Minnesota Press).

[8] For further details of these lists, see Western (note 2). There is also some discussion of them in Turner, B. A. (1976) *The Failure of Foresight* (Ph.D. thesis, University of Exeter) Ch. III.

[9] See Hamilton, J. A. B. (1967) *British Rail Accidents of the Twentieth Century* (London: Allen and Unwin).

[10] See Duckham, H. and Duckham, B. (1973) *Great Pit Disasters* (Newton Abbott: David and Charles).

[11] Western (note 2) draws attention to this discrepancy. See Montandon, F. (1959) Quelle est la moyenne annuelle des morts causés par les tremblements de terre? *Revue pour l'Etude des Calamités*, **36**, 61ff. and Keys (note 7).

[12] This approach is taken by Kröger, E. (1971) 'International assistance in natural disasters: experiences and proposals' (Diploma in Tropical Public Health, London School of Hygiene and Tropical Medicine, London University). See also Ambraseys, N. N. (Jan.—Feb. 1972) 'Earthquake hazards and emergency planning' *BUILD International*, 38–42.

[13] Western (note 2). A slightly different approach to this classification is suggested in Chapter 10.

[14] Western (note 2), p. 7.

[15] Farhar offers an account of a situation where there was a suspicion that floods had been produced by cloud-seeding, although they were eventually deemed not to be man-made. See Farhar B. C. (1976), The impact of the Rapid City flood on public opinion about weather modification, *Sociological Review*, 19 (No. 1), 117–143.

[16] Some of these points are relevant to the discussion of definitions of disasters in Chapter 5.

[17] Eddy, P. Potter, E. and Page, B. (1976) *Destination Disaster* (London: Hart Davis, MacGibbon). An earlier notable example of the genre is: Moscow, A. (1959) *Collision Course: The Story of the Collision between the* 'Andrea Doria' *and the* 'Stockholm' (London: Longmans).

[18] A recent review of these problems is provided in Richardson, J. W. ed. (1975) *Disaster Planning: Proceedings of a symposium held at the Royal Naval Hospital, Haslar, Gosport, Hants. on 10 and 11 October 1974* (Bristol: John Wright). On burns, see Batey, N. R. 'Summerland disaster'; 36–39 in these proceedings; on smoke inhalation, see Hampton, T. R. W. 'Maritime accidents: smoke inhalation injury' pp. 12–15. See also Hindle, J. F. 'Road traffic accidents: medical aspects' pp. 23–25; Rutherford, W. H. 'Civil disturbance' pp. 33–35; Lindsay, I. R. 'Nuclear accidents' pp. 30–32; and Mason, K. J. 'The aircraft accident' pp. 3–6, esp. p. 5.

[19] Some of the short papers in Richardson, J. W. (note 18) also provide an introduction to some of these medical-organizational problems. See, for example Bertram, J. 'Categorisation of casualties' pp. 52–54: Sillar, W. 'Reception and

triage of casualties' pp. 80–82: and Savage, P. E. A. 'Documentation' pp. 93–97.
[20] See Whitney, J. M. (1958) Disaster management: preventive medicine and public health *Military Medicine*, **123**, 181–188.
[21] Western, K. A. (note 2), p. 87.
[22] See Lorraine, N. S. R. (5 Feb. 1954) Canvey Island flood disaster, February 1954, *The Medical Officer*, **91**, 59–62.
[23] See Rennie, D. (3 Oct. 1970) After the earthquake, *The Lancet*, 704–707.
[24] See Orth, G. L. (1959) Disaster and disposal of the dead *Military Medicine*, **124**, 505ff.
[25] Western (note 2). See also the annotated bibliography produced by Manning, D. (1976) *Disaster Technology: an annotated bibliography* (Oxford: Pergamon).
[26] Saylor, L. F. and Gordon, J. E. (1957) The medical component of natural disasters *American J. of the Medical Sciences*, **234**, 342–362.
[27] Sommer, A. and Mosley, W. H. (13 May 1972) East Bengal cyclone of November 1970: epidemiological approach to disaster management *The Lancet*, 1029–1036. See also, Lechat, M. F. (1975). Disaster epidemiology, *International Journal of Epidemiology*, **4**, 5–7, though this blurs questions of the epidemiology of causes and of after-effects.
[28] See, for example: Australia. Royal Commission into the Failure of the West Gate Bridge *Report*, 1971 (Melbourne: Government Printer). This case is discussed in detail by Bignell (1977) in 'The West Gate Bridge collapse' in *Catastrophic Failures* ed. Bignell, V. Peters, G. and Pym, C. (Milton Keynes: Open University Press) pp. 127–165.
[29] As, for example Department of Employment (1975) *The Flixborough Disaster: Report of the Court of Inquiry* (London: HMSO).
[30] Feld, J. (1968) *Construction Failure* (New York: Wiley). For a similar but more recent account, see Geoff Scott (1975) *Building Disasters and Failures* (London: Construction Press).
[31] Feld, J. (note 30), pp. 17–20.
[32] See, for example, Campbell, J. (11 Aug. 1967) What happened to Apollo? *Space/Aeronautics Research and Development Handbook*. For a more general account of the mode of organization of the NASA administration, see Sayles, L. R. and Chandler, M. K. (1971) *Managing Large Systems: Organizations for the future* (New York: Harper & Row).
[33] See for example Otway, H. J. and Erdman, R. C. (1970) Reactor siting and design from a risk viewpoint, *Nuclear Engineering and Design* **13** (Pt 2), 365–368. A useful overview of the area is provided by Martin, J. (1976), Engineering reliability techniques, Unit 7/8, Course TD 342: *Systems Performance and Systems Failures* (Milton Keynes: Open University Press).
[34] A concise and illuminating discussion of the general principles lying behind the search for improved reliability is provided by Bompas-Smith, J. H. (1973) *Mechanical Survival: the use of reliability data* (New York: McGraw Hill). See also Martin J. (note 33). Other relevant issues are raised by Nixon, F. Frost, N. E. and Marsh, K. J. 'Choosing a factor of safety' pp. 136–140; and by Wise, S. 'Why metals break' pp. 125–135, both in Whyte, R. R. ed. (1975) *Engineering Progress Through Trouble: case histories drawn from the Proceedings of the Institution of Mechanical Engineers* (London: Institution of Mechanical Engineers).

[35] Lindquis, M. G. (1975) Analysis of system failure and corrective subsystems *Management Datamatics* **4,** (No. 1) 21–24, 22. Note that Lindquis excludes from his discussion totally unexpected failures of the kind which much of our later discussion will concentrate upon, although he does mention this kind of failure, referring to Platt's discussion of it which we shall encounter in Chapter 8.

[36] Brewer, R. (Nov. 1971) *The concept of failure with respect to the manufacture of electronic components.* Paper presented to Open University Seminar on Systems Failures, City University, London. Brewer is citing here an internationally accepted definition of failure in electronics equipment. On the design of 'disaster-resistant systems', see Rivas, J. R. and Rudd, D. F. (1975) Man-machine synthesis of disaster-resistant operations, *Operational Research* **23,** 2–21.

[37] See Lindquis, M. G. (note 35).

[38] This is true even in the routine process of quality control, where it is necessary to decide whether or not an item meets given quality control requirements: Turner, B. A. (1970) Control systems: development and interaction, 59–84 in Woodward, J. ed. *Industrial Organisation: Behaviour and Control* (Oxford: Oxford University Press).

[39] Lindquis, M. G. (note 35).

[40] See, in this context, Weick's discussion of prospective and retrospective rationalization. Weick, K. (1969) *The Social Psychology of Organizing* (Reading, Mass: Addison Wesley). Herbert Simon's notion of 'satisficing' discussed in Chapter 7 is clearly also relevant to these kinds of processes.

[41] Whyte, R. R. (note 34), p. 139.

[42] Whyte, R. R. (note 34), p. 1 and *passim*.

[43] Eyers, J. E. and Nisbett, E. G. 'Boilers' pp. 109–115 in Whyte, R. R. (note 34).

[44] For a popular account of these developments, see Rolt L. T. C. (1955) *Red for Danger: a history of railway accidents and railway safety precautions* (London: Lane, J.).

[45] See Dawson, P. H., Fidler, F. and Summers-Smith, D. '"Wire-wool" type journal failures' pp. 75–76; and p. 1. Jones *et al.* 'Development of large marine and high speed gearing' pp. 20–38, both in Whyte, R. R. (note 34).

[46] Dawson *et al.* (note 45), p. 76.

[47] Gibb, Sir C. 'Generator end-bell development: failures at Fulham and Richard Hearne' pp. 54–62 in Whyte, R. R. (note 34). Comment from p. 54.

[48] Birchon, D. 'The use and abuse of materials in ocean engineering' *ibid.* pp. 41–47.

[49] Coyle, M. B. and Watson, S. J. 'Castle Donington: a study of fatigue in turbine shafts' *ibid.* pp. 106–108.

[50] Whyte, R. R. 'Introduction' *ibid.*

[51] Scruton, C. and Coombs, T. A. 'The steel chimney vibration story' *ibid.* pp. 103–106.

[52] Two press items published in the same month indicate the wide range of possible unsuspected outcomes, one concerning the suspected induction of nickel carbonyl gas into an air-conditioning system: 'Mystery disease kills 19 after U.S. Convention' *Times* (London) 4 Aug. 1976, and subsequent accounts throughout August. The other concerns the unexplained growth of crystals in a hospital hot-water system: 'Doomwatch probe into mystery pipe crystals:

hospital water supply hit' *Birmingham Evening Mail* 7 Aug. 1976.
[53] Wise, S. 'Why metals break' pp. 125–135 in Whyte, R. R. (note 34).
[54] Birchon *op. cit.*
[55] Birchon *ibid.*
[56] Nixon, F., Frost, N. E. and Marsh, K. J. 'Choosing a factor of safety', pp. 136–140, in Whyte, R. R. (note 34).
[57] Whyte, R. R., 'Introduction', in Whyte, R. R. (note 34), p. 11.
[58] Havilland, Sir G. De and Walker, P. B., 'The Comet failure', pp. 51–53, in Whyte, R. R. (note 34), p. 53.
[59] One of these cases is the discussion by Fleeting, R. and Coats, R. of aspects of communication, administration and company policy which contributed to the failures of the turbine blades of the liner *Queen Elizabeth II:* 'The QE2 turbine blade failures' pp. 91–102 in Whyte, R. R. (note 34); and a second case is Sir Frank Whittle's account of the problems created for the development of the jet engine by the process of 'forced development' imposed upon it: 'The development of the Whittle jet engine' *ibid.*, pp. 3–12.
[60] For a statement of the case that science is not like this, see Feyerabend, P. (1973) *Against Method* (London: New Left Books).
[61] Comments made at the Open University Systems Failure Seminar, London, City University, November 1974.
[62] See Polanyi, M. (1958) *Personal Knowledge* (London: Routledge) and (1967) *The Tacit Dimension* (London: Routledge).
[63] For a discussion of such skills in connection with accidents, see Reason, T. (1977) Skill and error in everyday life, in Howe, M. J. A., ed., *Adult Learning: psychological research and applications* (London: Wiley) Chapter 2.
[64] This is argued clearly in relation to scientific training by Ravetz, J. R. (1971) *Scientific Knowledge and its Social Problems* (Oxford: Clarendon Press).
[65] Cresswell, H. B. (1943) *The Honeywood File* (London: Faber).
[66] Relevant developments which involve discussions of engineering and natural sciences such as geology are taking place with regard to the damage and disruption caused by earthquakes. The discussion of problems arising from goal-oriented systems created completely by engineers may be transferred without much difficulty to the information systems set up in order to try to provide advance warning of the behaviour of natural systems such as those which provoke earthquakes. The question of how far it is possible to predict such movements seems at the moment to be unresolved. See: D'Albe, M. F. (1966) Earthquakes: avoidable disasters, *Impact of Science on Society*, **16** (3) 189–202; (1970) Natural disasters: their study and prevention, *UNESCO Chronicle*, 16, 195ff; Haas J. E. and Ayre, R. S. (1969) *The West Sicily earthquake of 1968* (Washington D.C.: National Academy of Science/National Academy of Engineering). Ambraseys, N. N. (Jan.–Feb. 1972) Earthquake hazards and emergency planning, *BUILD International* 38–42. A useful introductory survey of studies of earthquakes and other geological hazards is provided by Bolt, B. A., Horn, W. L. Macdonald G. A. and Scott, R. F. (1975), *Geological Hazards: earthquakes, tsunamis, volcanoes, avalanches, landslides and floods* (New York: Springer Verlag).
[67] See, for example, the review of accident studies by Hale, A. R. and Hale, M. (1972) *A Review of the Industrial Accidents Research Literature* (London: HMSO).

[68] See Robens Report *Safety and Health at Work* (1972) Cmnd 5034 (London: HMSO); and Health and Safety Commission *First Report of the Advisory Committee on Major Hazards* (1976) (London: HMSO). For a critical review of the assumptions underlying the safety at work legislation, see Nichols, T. and Armstrong, P. (1973) *Safety or Profit: Industrial accidents and the conventional wisdom* (Bristol: Falling Wall Press).

[69] Approximately 1000 people are killed at their work every year in Britain, according to the Robens Report (note 68), para. 10, while 7779 people were killed on the roads in Britain in 1972. *Annual Abstract of Statistics* (1973) (London: HMSO).

[70] See Chapanis, A. (1959) *Research Techniques in Human Engineering* (Baltimore: Johns Hopkins Press).

[71] See, for example, Macmonagle, L. N. (Oct. 1973) Accident prevention in rail transit, *Proceedings of the 17th Annual Meeting of the Human Factors Society*, pp. 483–496. See also Martin, J. (note 33).

[72] See, for example, Goeller, B. F. (1969), Modelling the traffic safety system, *Accident Analysis and Prevention*, **1,** 167–204; Hale, A. H. and Hale, M. (1970) Accidents in perspective, *Occupational Psychology*, **44,** 115–121; Wigglesworth, E. L. (1972) A teaching model of injury causation and a guide for selecting countermeasures. *Occupational Psychology,* **46,** 69–78; and Lawrence, A. C. (1974) Human error as a cause of accidents in gold-mining, *Journal of Safety Research*, 6, 78–88.

[73] Nichols, T. and Armstrong, P. (note 68), p. 5. Nichols and Armstrong refer specifically to the social relations of production affecting industrial accidents, but it seems to me that their criticism could be broadened to include not only factors of class and industrial power, but a range of other social structural characteristics which are commonly neglected in the discussion and analysis of accidents. It should also be noted, however, that as Nichols and Armstrong comment (p. 5) industrial sociologists who might have been expected to compensate for this neglect have not, in fact, done so (though see note 75).

[74] Lawrence, A. C. (note 72). The absence of comment on the influence of social structural factors is particularly noticeable in a study based on the South African gold-mining industry, however.

[75] As noted in note 73, sociologists do not appear to have been particularly active in the study of industrial accidents, although one or two brief contributions may be noted. Eldridge, J. E. T. and Kaye, B. M. have discussed some of the methodological problems which arise in studying the causation of accidents, relating them to the intended and unintended consequences of social and technical norms 'Wages and accidents; an exploratory paper' in Teulings, A. ed. (1973) *Ondernemingen vakbeweging* (The Hague: Mens en Maatschappij) pp. 152–171. Baldamus and his colleagues have reviewed a number of aspects of accidents, wrestling, among other things, with the somewhat intractable problem of accounting for the persistent variations in accident statistics according to days of the week. See: Baldamus W. (1969) *Alienation, Anomie and Industrial Accidents* Discussion Papers Series E, No. 12 (Birmingham: Faculty of Commerce, University of Birmingham, duplicated). Baldamus W. (1969) *The Concept of Truly Accidental Accidents* Discussion Papers Series E No. 14. (Birmingham: Faculty of Commerce, University of Birmingham, duplicated). See also the useful list of references provided in Baldamus W. (1976) Krankenstand,

Fehlzeiten und Arbeitsunfälle: Internationaler Vergleich, in *Handbuch der Sozialmedizin* (3 vols.) ed. Blohmke, M. v. Ferber, C., Kisker, K. P. and Schaefer, H. (Stuttgart: Ferdinand Enke Verlag) Vol. III pp. 88–99.

[76] Robens Report (note 68).

[77] See, for example, Healy, R. J. (1969) *Emergency and Disaster Planning* (New York: Wiley).

[78] See Saylor and Gordon (note 26).

[79] Richwagen, W. C. (6 Aug. 1967) The predictive approach to disaster planning—how it failed, *Hospitals: Journal of the American Hospitals Association*, **41**, 48–51. The conceptual problem underlying Richwagen's account is very similar to what has been called the 'noisy marbles' problem. That is, the difficulty of classifying items into a given scheme as more 'noisy' features are added to the items. Macey, J. Christie and Luce (1953) Coding noise in a task-oriented group, *Journal of Abnormal and Social Psychology*, **48**, 401–409. See also Brown, R. (1965) *Social Psychology* (New York: Free Press) pp. 332–349.

[80] *The Flixborough Disaster* (note 29). The increased number of articles on safety topics and, indeed, the recent crop of new journals in the area, offer some indicators of a greater awareness of the problems of safety. New journals in addition to more established publications such as the *Journal of Safety Research* include the *Journal of Hazardous Materials* (1975) *Mass Emergencies* (1975) and *Disasters* (1977). For a discussion of safety problems from a personnel angle see Gill, J. and Martin, K. (June 1976) Reconciling rules with reality, *Personnel Management*, 36–39. A somewhat superficial account of the policy-making aspects of hazard control is presented in Chicken, J. C. (1975) *Hazard Control Policy in Britain* (Oxford: Pergamon).

[81] Starr, C. (1969), Social benefit versus technological risk: what is our society willing to pay for safety. *Science*, **165**, 1232–1238. For a discussion of some of the problems of decision-making with regard to hazards at a national level, see Dunkelman, H. (1976) *Science Policy*, Unit 9/10. Course TD342 Systems Performance: Human Factors and Systems Failures (Milton Keynes: Open University Press).

[82] See Chicken (note 80) and Martin (note 33), p. 63.

[83] Bowen, J. H. (19 Oct. 1968), Risk from a supernova compared with the risk standard for nuclear reactors, *Nature*, **220**, 303–304.

[84] On nuclear engineering, see Martin (note 33); and Otway and Erdman (note 33). An account of the ICI approach to risks from accidental chlorine emissions is given by Dicken, A. N. A. (May 1974) The quantitative assessment of chlorine emission hazards, in *Report of the Bi-centennial Chlorine Symposium* (San Francisco: The Electro-Chemical Society and Chlorine Institute) pp. 244–256.

[85] On forecasting, see Encel, S. Marstrand, P. K. and Page, W. (1975) *The Art of Anticipation: values and methods in forecasting* (London: Martin Robertson). For an interesting discussion of risk in a commercial–historical setting, see Strassman, R. (1959) *Risk and Technological Change* (New York: Cornell University Press), esp. Chapter 1.

[86] See Williams, P. (1975) *Crisis Management* (London: Martin Robertson); Stallings, R. A. (1975) Differential response of hospital personnel to a disaster, *Mass Emergencies*, **1** (No. 1), 47–54; and Egan, W. (Jan. 1976) Crisis management: a topic for the future, paper delivered to the Conference of the Oc-

cupational Psychology Section of the British Psychological Society.
[87] By Pettigrew, A. (Autumn 1970) Learning from extreme situations, *London Business School Journal*, **4**, 18–20, 29.
[88] See Fink, S. L., Beak, J. and Taddeo, K. (1971) Organizational crisis and change, *Journal of Applied Behavioral Science*, **7** (No. 1), 15–37.
[89] Zaleznik, A. (1967) Management of disappointment, *Harvard Business Review*, **45**, 59–70.
[90] At least, so Sister Marie Augusta Neal, SND (1965) assures us: *Values and Interests in Social Change* (Englewood Cliffs, NJ: Prentice Hall) p. 8.
[91] Egan (note 86).
[92] Hughes, E. C. (1958) discusses this aspect of professional occupational behaviour, among many others, in *Men and Their Work* (Glencoe Ill: Free Press).
[93] On crises and growth, see Lippitt, G. L. and Schmidt, W. H. (1967) Crises in a developing organization, *Harvard Business Review*, **45**, 102–112. On planning failures, see Schaffer, R. H. (1967) 'Putting action into planning' *Harvard Business Review*, **45**, 158–166. And on merger failures, Levinson, H. (1970) A psychologist diagnoses merger failures, *Harvard Business Review*, **48**, 139–147.
[94] See Schon, D. A. (1971) *Beyond the Stable State: Public and Private Learning in a Changing Society* (London: Temple Smith).
[95] Considerable steps towards such an approach, within a systems framework, have been made by the various contributors to the course units on Open University course TD 342 *Systems Performance: Human Factors and Systems Failures*, in addition to those by Martin and Dunkelman cited above. See also *Catastrophic Failures* (note 28).

Chapter 3

[1] Prince, S. H. (1920) *Catastrophe and Social Change: based upon a sociological study of the Halifax disaster* (New York: Columbia University Press); Queen, S. A. and Mann, D. M. (1925) *Social Pathology* (New York: Thomas Y. Crowell); Carr, L. J. (1932), Disaster and the sequence-pattern concept of social change, *American Journal of Sociology*, **38**, 207–218. With regard to the subsequent discussion, see the comment on Carr in note 7.
[2] Kastenbaum, R. (1974) Disaster, death and human ecology, *Omega, Journal of Death and Dying*, **5** (No. 1), 65–75.
[3] The 'restorative' emphasis of Prince's study, for example, comes through clearly in his comment in the Preface that: 'This awful catastrophe is not the end but the beginning. History does not end so. It is the way its chapters open.' Further on, in the Introduction, he comments that by shattering the 'matrix of custom' catastrophe may even spur on progress. Prince (note 1).
[4] Prince (note 1).
[5] It is interesting, in this context, to note that Williams, R. M. and Parkes, C. M. (1975) have shown a similar compensatory effect in the movements of the birth rate in Aberfan since the occurrence of the disaster there: 'Psychosocial effects of disaster: birth rate in Aberfan' *British Medical Journal*, **2**, 303–304.
[6] Prince (note 1). Chapter I.
[7] Carr (note 1). It is true that Carr does include as the first stage of his

sequence model a preliminary or 'prodromal' phase, which includes specifically, in his discussion of Prince's study, the twenty minutes or so before the collision when the two ships were steaming towards each other, but this phase is treated in a very cursory manner and does not shift the general emphasis from the post-disaster stages.

[8] See Baker, G. W. (1964) Comments on the present status and future directions of disaster research, in Grosser, G. H., Wechsler, H. and Greenblatt, M. eds. *The Threat of Impending Disaster: Contributions to the Psychology of Stress* (Cambridge, Mass: MIT Press), pp. 315–330.

[9] Wallace mentions this as a source of medical interest in the area. Wallace, A. F. (1956) *Tornado in Worcester: An exploratory study of individual and community behavior in an extreme situation* Committee on Disaster Studies, Study No. 3. Publication 392, (Washington D.C.: National Academy of Science/National Research Council).

[10] See, for example, Hersey, J. (1946) *Hiroshima* (New York: Knopf); and Lifton, R. J. 'Psychological effects of the atomic bomb in Hiroshima: the theme of death' in Grosser, Wechsler and Greenblatt, eds. (note 8), pp. 152–193.

[11] As, for example, in the writings of Bahnson, C. B. 'Emotional reactions to internally and externally derived threats of annihilation' in Grosser, Wechsler and Greenblatt, eds. (note 8), pp. 251–280; Betelheim, B. (1943) Individual and mass behavior in extreme situations, *Journal of Abnormal and Social Psychology*, **39**, 417–452; and Biderman, A. D. 'Captivity lore and behavior in captivity' in Grosser, Wechsler and Greenblatt, eds. (note 8), pp. 223–250.

[12] Killian, L. M. with the assistance of Quick, R. and Stockwell, F. (1956) *A Study of Response to the Houston, Texas, Fireworks Explosion*, Committee on Disaster Studies. Disaster Study No. 2. Publication No. 391. (Washington D.C.: National Academy of Sciences/National Research Council). The NAS/NRC programme which this study formed a part of, and which continued until 1963, was itself an outgrowth and continuation of the wartime research effort of the 'American Soldier' research programme, directed by S. Stouffer.

[13] For a discussion of similar media exaggerations of response to news of a major accident, in this case a nuclear one, see Rosengren, K. E., Arvidson, P. and Sturesson, D. (1975), The Barsebäck panic: a radio programme as a negative summary event, *Acta Sociologica*, **18** (No. 4), 303–321.

[14] The equation of the problems of disaster with the problems of civil defence in the case of a nuclear attack is clearly expressed, for example, in Marden's short discussion of contributions to the field: Marden, R. H. (1960) Disaster! *Public Administration Review*, **20**, 100–105.

[15] See, for example, Williams, H. B. (1957) Some functions of communication in crisis behavior, *Human Organization*, 16 (No. 2), 15–59; Mack, R. W. and Baker, G. W. (1961) *The Occasion Instant: the structure of social responses to unanticipated air-raid warnings* Disaster Study No. 15. (Washington D.C., National Academy of Sciences/National Research Council); Withey, S. B. 'Sequential accommodations to threat' in Grosser, Wechsler and Greenblatt, eds. (note 8), pp. 105–114. Williams, H. B. 'Human factors in warning-and-response systems' pp. 79–104 *ibid.*; and Anderson, W. A. (1969) Disaster warning and communication processes in two communities, *Journal of Communication*, **19**, 92–104. Williams' discussion draws attention to an interesting journalistic account of some of the problems involved in deciding

just when to activiate a warning system: Brooks, J. (28 May 1955) Reporter at large: five-ten on a hot sticky day, *New Yorker* pp. 39ff.

[16] Grinspoon, L. 'Fallout shelters and the acceptability of disquieting facts' in Grosser, Wechsler and Greenblatt, eds. (note 8), pp. 117–130.

[17] Larsen, O. N. (1954) Rumors in a disaster, *Journal of Communications*, **4**, 111–123; Wenger, D. E. and Parr, A. (1969) *Community Functions under Disaster Conditions* Report No. 4. (Ohio: Ohio State University Disaster Research Center).

[18] See, for example, Chapman, D. W. (1954) Human behavior in disaster: a new field of social research, *Journal of Social Issues*, **10** (No. 3), 1–73. For a more recent discussion of some aspects of threat, see Wolf, C. (Summer 1975), Group perspective formation and strategies of identity in a post-threat situation, *Sociological Quarterly*, **16**, 404–414.

[19] See, for example, Clifford, R. A. (1956) *The Rio Grande Flood: a comparative study of border communities in disaster* Disaster Study No. 7 (Washington D. C., National Academy of Sciences/National Research Council); Wallace, F. C. (note 9); and Burton, I., Kates, R. W. and White, G. F. (1968) *The Human Ecology of Extreme Geophysical Events*, Natural Hazards Research Working Paper No. 1 (Toronto, Canada: University of Toronto).

[20] Thompson, J. D. and Hawkes, R. W. (1962) Disaster, community organization and administrative process, in *Man and Society in Disaster*, eds. Baker, G. W. and Chapman, D. W. (New York: Basic Books); Kennedy, W. C. (1970) Police departments: organization and tasks in disaster, *American Behavioral Scientist*, **13** (No. 3), 354–361; Warheit, G. J. 'Fire department's operations during major community emergencies' *ibid.* pp. 362–368. For a criticism of this general approach, see Maley, R. F. (1974) Conceptual disasters about disasters, *Omega, Journal of Death and Dying*, **5** (No. 1), 73ff.

[21] This topic is reviewed in Fritz, C. E. and Mathewson, J. H. (1957) *Convergence Behavior in Disasters: a problem of social control*, Disaster Study No. 9 (Washington D.C.: National Academy of Sciences/National Research Council).

[22] See, for example, Roth, R. (1970) Cross-cultural perspectives on disaster responses, *American Behavioral Scientist*, **13** (No. 3), 440–451. A recent exception to this general approach is provided by Oliver-Smith, A. (Spring 1977) Disaster rehabilitation and social change in Yungay, Peru, *Human Organization*, **36** (No. 1), 5–13.

[23] Schneider, D. M. (1957) Typhoons on Yap, *Human Organization*, **16** (No. 2), 10–15. Quotation p. 14.

[24] Kastenbaum, R. (note 2). Quotation p. 65.

[25] An interesting interpretation of the relationship between certain types of event such as accidents, and the response which they provoke in the media is provided by Rosengren *et al.* (note 13) in their discussion of 'summary events' and 'negative summary events'. On media response to accidents, see also Molotch, H. and Lester, M. (1975) Accidental news: the great oil spill as local occurence and national event, *American Journal of Sociology*, **81** (No. 2), 235–260.

[26] Western (chap. 2, note 2), p. 1.

[27] Western (chap. 2, note 2) p. 1.

[28] Western (chap. 2, note 2) *passim.*; Maley (note 20); Kastenbaum (note 2):

Turner, B. A. (1976) *The Failure of Foresight* (Ph.D. thesis, University of Exeter).

[29] Baker, G. W. 'Comments on the present status and future directions of disaster research' in Grosser, Wechsler and Greenblatt, eds. (note 8), pp. 315–330.

[30] Dynes, R. R. (1973) *Organized Behavior in Disaster* (Lexington, Mass.: Heath Lexington Books). The post-disaster perspective is evident, too, in the publications and the kinds of research supported in the United Kingdom by the Action For Disaster fund which was set up in Scotland following the Ibrox Park football crowd disaster. See Action For Disaster (1973) *A Guide to Disaster Management* (Glasgow: Action For Disaster); and (1976) *Developments in Disaster Management* (Glasgow: Action For Disaster). And, again, the same emphasis is evident throughout the various contributions to the symposium on disaster planning reported in Richardson, J. W. Ed. (1975) *Disaster Planning: Proceedings of a symposium held at the Royal Naval Hospital, Haslar, Gosport, Hants, on 10 and 11 October 1974* (Bristol: John Wright): and also in Green, S. (1977) *International Disaster Relief: toward a responsive system* (New York: McGraw Hill).

[31] Drabek, T. E. (1970) Methodology of studying disasters: past patterns and future possibilities, *American Behavioral Scientist*, **13** (No. 3), 331–343.

[32] Powell, J. W. Rayner, J. and Finesinger, J. E. (1953) Responses to disaster in American cultural groups, in *Symposium on Stress: 16–18 March 1953*. Sponsored jointly by Division of Medical Sciences, National Research Council and the Army Medical Service Graduate School, Walter Reed Army Medical Center. (Washington D.C.: Army Medical Service Graduate School) pp. 174–193.

[33] Barton, A. H. (1969) *Communities in Disaster: a sociological analysis of collective stress situations* (London: Ward Lock) p. 53. The phases which Barton discerns in his concern to study 'collective stress situations' are: 1. Predisaster; 2. Detection and communication of warnings of a specific threat; 3. Immediate, relatively unorganized social response; 4. Organized social response; 5. Long-run post-disaster equilibrium, when reconstruction is completed. Other 'stage' formulations organized around the moment of impact are those presented by Wallace (note 9) and Carr (note 1).

[34] Wallace (note 9).

[35] Taylor, I. and Knowlenden, J. (1957) *Principles of Epidemiology* (Boston: Little, Brown).

[36] As, for example, in Ennis, J. (1959) *The Great Bombay Explosion* (New York, Duell, Sloan and Pearce); in Grieve, H. (1959) *The Great Tide* (Essex, Essex County Council); or in Moscow, A. (1959) *Collision Course: the story of the collision between the 'Andrea Doria' and the 'Stockholm'* (London: Longmans).

[37] Wilensky, H. L. (1967) *Organizational Intelligence* (New York: Basic Books).

[38] Wohlstetter, R. (1962) *Pearl Harbor: Warning and decision* (Stanford: Stanford University Press). In many ways, the scale and complexity of the intelligence build-up to Pearl Harbor which Wohlstetter describes, and the number of clues available to the Americans, make this situation comparable to the complex interplay of events which often precedes inadvertent disasters. In order to draw such a parallel, we need to see the Japanese as equivalent in some way to

the hidden forces of nature, or, more appropriately, to regard nature as a cunning adversary. It should be noted, however, that Wohlstetter's book presents a detailed and scholarly account of the intelligence situation before Pearl Harbor from the point of view of a political scientist interested in military history. She uses the mass of detailed information that she assembles to offer some low-level generalizations about the Pearl Harbor incident, and its implications for American defence policies in the future, but the wider discussion of Pearl Harbor as a case study in the failure of foresight is left entirely to Wilensky.

[39] Hudson, B. B. (1954) Anxiety in response to the unfamiliar, *Journal of Social Issues*, **10** (No. 3), 53–60.

[40] Janis, I. L. 'Psychological effects of warnings' in Baker and Chapman, eds. (note 20).

[41] Lang, K. and Lang, G. E. 'Collective responses to the threat of disaster' in Grosser, Wechsler and Greenblatt, eds. (note 8), pp. 58–75.

[42] A systems model is proposed both by Stanton, A. H. 'Situations evoking stress in human groups and the group behavioral changes denoting strain' in *Symposium on Stress* (note 32), pp. 165–174; and by Miller, J. C. 'A theoretical review of individual and group psychological reactions to stress' in Grosser, Wechsler and Greenblatt, eds. (note 8) pp. 11–33.

[43] Miller (note 42) regards the problem as one of 'input information overload'. See also the discussion of 'variable disjunction of information' by Reeves, T. K. and Turner, B. A. (1972) A theory of organization and behavior in batch production factories: *Administrative Science Quarterly*, **17**, 81–98. See also the discussion of this concept in the following chapter.

[44] See Ruesch, J. 'The interpersonal communication of anxiety' in *Symposium on Stress* (note 32), pp. 154–164; and Speigel, J. P. 'Psychological transactions in situations of acute stress', *ibid.*, pp. 103–115.

[45] Demerath, N. J. (1957) Some general propositions: an interpretative summary, *Human Organization*, **16** (No. 2), 28ff.

[46] Kilpatrick, F. P. (1957) Problems of perception in extreme situations, *Human Organization*, **16** (No. 2), 20ff.

[47] This problem is discussed in the context of accidents in chemical plants by Rivas, J. R. and Rudd, D. F. (1975) Man–machine synthesis of a disaster-resistant system, *Operational Research*, **23** (No. 1), 2–21.

[48] Wolfenstein, M. (1957) *Disaster: a psychological essay* (London: Routledge and Kegan Paul). Quarantelli is unduly critical and dismissive of Wolfenstein's insightful and rewarding discussion in his comments in Quarantelli, E. L. (1970) A selected annotated bibliography of social science studies on disasters, *American Behavioral Scientist*, **13** (No. 3), 452–456.

[49] This sequence is suggested by Williams, H. B. 'Human factors in warning-and-response systems' pp. 79–104 in Grosser, Wechsler and Greenblatt, eds. (note 8).

[50] Brooks, J. (note 15).

[51] Anderson, W. A. (note 15).

[52] *ibid.*

[53] Fritz, C. E. and Williams, R. B. (1957) The human being in disasters: a research perspective, *Annals of the American Academy of Political and Social Science*, **309**, 42–51.

[54] Withey, S. B. 'Sequential accommodations to threat' in Grosser, Wechsler

and Greenblatt, eds. (note 8), pp. 105–114.
[55] See Fritz and Williams (note 53).
[56] See Williams (note 49).

Chapter 4

[1] Much of the material presented in this chapter was published, in a slightly modified form, in Turner, B. A. (1976) An initial analysis of Hixon, Aberfan and Summerland, *Catastrophe and its preconditions* ed. Peters, G. *et al.* Unit 4 of Open University Systems Performance Course TD 342 (Milton Keynes: Open University Press), pp. 15–44; and some of the material has been incorporated into Turner, B. A. (1976) The organizational and interorganizational development of disasters, *Administrative Science Quarterly*, **21**, 378–397.
[2] See Turner, B. A. (1970) The organization of production scheduling in complex batch production situations, in *Approaches to Organizational Behavior*, ed. Heald, G. (London: Tavistock) pp. 87–99; and Reeves, T. K. and Turner, B. A. (1972) A theory of organization and behaviour in batch production factories, *Administrative Science Quarterly*, **17**, 81–98.
[3] Wilensky, H. L. (1967) *Organizational Intelligence* (New York: Basic Books).
[4] For a more detailed discussion of the method used in the present investigation, see Turner, B. A. (1976) *The Failure of Foresight* (Ph.D. thesis, University of Exeter).
[5] In fact, in the present case, alternative interpretations of some of these initial reports have been produced, with public discussion and comparison of these and the present studies, but although the interpretations are not identical, there seems to have been some agreement that they are not conflicting, the differences merely reflecting the different disciplines within which interpretations have been constructed, and the different concerns of the investigators. See Spear, R. (1976) The Hixon Analysis, *Unit 2/3* Open University Systems Performance Course TD342 (Milton Keynes: Open University Press); and Bignell, V. (1976) Case Study: Hixon level crossing accident, in *Course Reader 1*, Open University Systems Performance Course TD 342, CR 1. (Milton Keynes: Open University Press).
[6] See Turner, B. A. (note 2) and Reeves, T. K. and Turner, B. A. (note 2).
[7] Devons has described much the same problem and much the same solution arising in the wartime planning activities of the Ministry of Aircraft Production. Devons, E. (1950) *Planning in Practice: essays in aircraft planning in wartime* (Cambridge: Cambridge University Press).
[8] The terms 'well-structured problem' and 'ill-structured problem' are used by Tonge, F. M. (1961) *A Heuristic Program for Assembly Line Balancing* (Englewood Cliffs, N.J. Prentice Hall).
[9] See, for example, *Report of the Inquiry into the Fire at Micheal Colliery, Fife* (1967) Cmnd 3657 (London, HMSO); and Ministry of Power *Report of the Inquiry into the Causes of the Accident to the Drilling Rig, Sea Gem*, Cmnd 3409 (London: HMSO).
[10] House of Commons Paper, *Report by the Tribunal of Inquiry* (1966–67), HC 553 (London: HMSO).
[11] *Report of the Public Inquiry into the accident at Hixon Level Crossing on 6*

January 1968 (1968) Cmnd 3706 (London: HMSO).
[12] Summerland Fire Commission *Report of the Summerland Fire Commission*, undated—1974 (Isle of Man: Government Office).
[13] More extended summaries and discussions of the three incidents are available in Bignell, V. Peters, G. and Pym, C. (1977) *Catastrophic Failures* (Milton Keynes: Open University Press) Chapters 1–3.
[14] Summerland para. 246. References to the three reports in the remainder of this chapter will be given as paragraph references to the 'Summerland', 'Hixon' or 'Aberfan' reports. A complete list of public inquiry reports consulted in the course of the investigation will be found in the Appendix, pp. 000–000.
[15] See Turner, B. A. (1971) *Exploring the Industrial Subculture* (London: Macmillan).
[16] Aberfan, para. 15.
[17] Whorf, B. L. (1956) The relation of habitual thought and behavior to language, in *Language, Thought and Reality*, ed. Carroll, J. B. (Cambridge, Mass.: MIT Press).
[18] Summerland, para. 60.
[19] Summerland, para. 206.
[20] Aberfan, para. 17.
[21] Aberfan, paras. 96, 99.
[22] Aberfan, paras. 101–103.
[23] Aberfan, para. 72.
[24] Aberfan, para. 72.
[25] Discussions of vicious circles as characteristic bureaucratic phenomena will be found in Crozier, M. (1964) *The Bureaucratic Phenomenon* (London, Tavistock); and in Gouldner, A. (1954) *Patterns of Industrial Bureaucracy* (New York: Free Press).
[26] Aberfan, paras. 123, 127, 159–172.
[27] Aberfan, para. 147.
[28] Hixon, para. 112.
[29] Hixon, paras. 125, 130.
[30] Hixon, para. 161.
[31] Hixon, para. 162.
[32] Aberfan, paras. 64, 92, 146, 149.
[33] Hixon, paras. 109, 113.
[34] For a discussion of this option, see Meier, R. L.]1965) Information input overload: features of growth in communications oriented institutions, in *Mathematical Explorations in Behavioral Science*, ed. Massarik, F. and Ratoosh, P. (Homewood, Ill: Irwin, R. D.) pp. 233ff.
[35] Hixon, para. 221.
[36] Summerland, para. 1.5.
[37] Summerland, paras. 1.3, 4.2.
[38] Summerland, para. 5.3.
[39] Aberfan, para. 14.
[40] Summerland, paras. 1.4, 4.1, 4.2.
[41] Summerland, para. 161. See also para. 4.2.
[42] Hixon, para. 7.1.
[43] Hixon, para. 176.
[44] Summerland, para. 5.1.

[45] Aberfan, para. 9.
[46] Hixon, para. 10.
[47] Summerland, para. 1.4.
[48] Hixon, para. 9.2. Compare also Summerland, para. 5.2.
[49] Summerland, para. 1.4.
[50] Hixon, para. 193.
[51] Summerland, para. 4.1.
[52] Hixon, para. 9.2.
[53] Summerland, para. 4.2.
[54] Summerland, para. 2.3.
[55] Hixon, para. 10.
[56] Hixon, para. 192.
[57] Hixon, para. 193.
[58] Hixon, para. 184.
[59] Summerland, para. 5.2.
[60] See, for example, Polya, G. (1944) *How to Solve It* (Princeton: Princeton University Press); Steiner, G. A. ed. (1965) *The Creative Organization* (Chicago: University of Chicago Press); and Perloff, H. S. (1974) Knowledge to action: creating an undergraduate problem-solving program, *American Behavioral Scientist*, **18,** 211–311.
[61] Wolfenstein, M. (1957) *Disaster: a Psychological Study* (London: Routledge and Kegan Paul).
[62] Summerland, para. 4.2.
[63] Hixon, paras. 6.2, 7.2.
[64] Hixon, para. 6.4.
[65] Hixon, para. 6.3.
[66] Summerland, para. 4.2.
[67] See Pask, G. (1960) The natural history of networks, in *Self-Organizing Systems*, ed. Yovits, M. C. and Cameron, S. (London: Pergamon).
[68] The interrelationships of system properties are discussed at a number of points by Pask. See Pask (note 67); and also (1960) Teaching machines, in *Proceedings of the 2nd International Congress of Cybernetics, Namur, 1958* (Association Internationale de Cybernetique) pp. 962–978; and (1962) in Interaction between a group of subjects and an adaptive automation (*sic*) to produce a self-organizing system for decision-making, in *Self-Organizing Systems 1962*, eds. Yovits, M. C., Jacobi, G. T. and Goldstein, G. D. (Washington: Spartan Books).
[69] Summerland, para. 4.1.
[70] For example, it was not realized that the control room at Summerland was also an emergency communications centre. Summerland, para. 4.2.
[71] Wolfenstein, M. (note 61).
[72] Succinct summary definitions of the concepts of 'stranger' and 'site' set out here have been formulated by the Open University Systems Performance team:

> *Strangers* are people who have access to a part of a system, not necessarily legally, and who cannot be adequately briefed about the situation, because they are not sufficiently clearly identified to enable training to take place, or because the 'keepers of the system' do not have sufficient influence over them to require them to be informed. A *site* is a concrete sub-system, the components of which have additional properties to those required for the system.

Peters, G. *et al.*, eds. (note 1), p. 45.

[73] Summerland, para. 5.2.
[74] Summerland, para. 1.3.
[75] Summerland, para. 1.4.
[76] Summerland, paras. 2.4, 4.
[77] Hixon, para. 11.
[78] Summerland, para. 1.4.
[79] Summerland, para. 1.4.
[80] Summerland, para. 1.4.
[81] Summerland, para. 1.4.
[82] Summerland, para. 1.4.
[83] Summerland, para. 1.4.
[84] Summerland, paras. 1.4, 5.5.
[85] Summerland, para. 4.1.
[86] Summerland, para. 5.3.
[87] Summerland, para. 5.3.4.
[88] Summerland, paras, 1.4, 5.1, 5.5.
[89] Wolfensteins' argument, from a psychoanalytic point of view, that the normal response to danger is to deny it, seems to be relevant here as offering a possible form of explanation. The individual is likely to persist in this denial until he experiences a 'subjective near-miss', an experience in which his own vulnerability is brought directly home to him. Wolfenstein in her extended discussion of denial, in Wolfenstein, M. (note 61), chap. III, includes under this heading: (a) low cathexis—slight or rare concern with an unpleasant fact, which is mainly disregarded, (b) repression, (c) isolation of affect from cognitive indications of danger, and (d) intellectual but not behavioural acknowledgment of dangerous situations.
[90] Aberfan, para. 11.
[91] Aberfan, para. 9.
[92] Hixon, para. 10.
[93] Summerland, para. 185. See also Wolfenstein's discussion of propitiation. Wolfenstein, M. (note 61), Chap. IV.
[94] Aberfan, para. 10.
[95] Hixon, para. 7.3.
[96] For discussions of rationalizations and other reactions to the interplay of ideologies, see Berger P. and Luckman, T. (1967) *The Social Construction of Reality* (London: Allen Lane); and Turner, B. A. (1971) *Exploring the Industrial Subculture* (London: Macmillan) pp. 114–119.
[97] Aberfan Report, Table I.
[98] Aberfan, para. 8.
[99] Aberfan, para. 9.
[100] Aberfan, para. 197.
[101] Hixon, para. 114.
[102] Hixon, para. 7.3.
[103] Aberfan, para. 10; Hixon, para. 8; Summerland, paras. 4.1, 4.2, 5.2.
[104] See, for example, Aberfan, paras. 60, 93, 96, 105, 174–176; or Summerland paras. 188 and 219.
[105] Wolfenstein, M. (note 61).
[106] Aberfan, para. 10.
[107] Hixon, para. 7.3.

[108] Summerland, paras. 4.1, 4.2.
[109] Barlay, S. (1972) *Fire: an international report* (London: Hamish Hamilton).
[110] Wolfenstein, M. (note 61).
[111] Summerland, para. 5.9.
[112] In sociological terms, the array of factors set out may be considered to be an 'ideal type' in the rather specialized sense in which this term has been used by M. Weber. The discussion is therefore intended to present a compilation of characteristics which represent the central features of the material studied up to this point, and which are compatible with each other so that, conceivably, they could occur together. The main purpose of an ideal-type is to crystallize such features as have been perceived in earlier work, in order to aid future analysis.
[113] This, among other possibilities, is discussed in Ravetz, J. R. (1974) Comment: the safety of safeguard, *Minerva*, **XII**, 323–326.
[114] Compare the strategy adopted in Reeves, T. K. and Turner, B. A. (note 1).

Chapter 5

[1] Some of the material in this chapter has already appeared in a slightly modified form in Turner, B. A. 'The development of disasters' *Sociological Review*, **24** (No. 4, 1976), 753–774.
[2] The reports used were: *Explosion at Cambrian Colliery, Glamorgan* (1966) Cmnd 2813 (London: HMSO). House of Commons Paper *Report by the Tribunal of Inquiry (Aberfan)* (1966–67) HC 553 (London: HMSO). *Report of the Inquiry into the Causes of the Accident to the Drilling Rig, Sea Gem* (1967) Cmnd 3409 (London: HMSO). *Report of the Inquiry into the Fire at Micheal Colliery, Fife* (1968) Cmnd 3657 (London: HMSO). *Report of the public Inquiry into the Accident at Hixon Level Crossing on 6 January 1968* (1968) Cmnd 3706 (London: HMSO). *Report of the Committee of Inquiry into the Fire at Coldharbour Hospital, Sherborne on 5 July 1972* (1972) Cmnd 5170 (London: HMSO). *Report by HM Inspectorate on the Collapse of Falsework for the Viaduct over the River Loddon on 24 October 1972 (1973)* House of Commons paper HC 425 (London: HMSO). *Inrush at Lofthouse Colliery, Yorkshire* (1973) Cmnd 5419 (London: HMSO). *Extensive Fall of Roof at Seafield Colliery, Fife* (1973) Cmnd 5485 (London: HMSO). *Accident at Markham Colliery, Derbyshire* (1974) Cmnd 5557 (London: HMSO). *Report of the Committee of Inquiry into the Smallpox Outbreak in London in March/April 1973* (1974) Cmnd 5626 (London: HMSO). Summerland Fire Commission *Report of the Summerland Fire Commission*, undated—1974 (Isle of Man: Government Office). Department of Employment (1975) *The Flixborough Disaster: Report of the Court of Inquiry* (London: HMSO).
   These incidents led to a total of 351 deaths, with the number of deaths attributable to a single incident varying from 2 to 144.
[3] The discussion in this section has been considerably aided by the useful review provided by Western, K. A. (Chap. 2, note 2), a review which we have already drawn upon extensively in Chaps. 2 and 3.
[4] By Quarantelli, E. L. and Dynes, R. R. (1970) Editorial introduction,

American Behavioral Scientist, **13** (No. 3), 325–330.
[5] Saylor, L. F. and Gordon, J. E. (1957) The medical component of natural disasters, *American Journal of the Medical Sciences*, **234**, 342–362.
[6] Kastenbaum, R. (1974) Disaster, death and human ecology, *Omega, Journal of Death and Dying*, **5** (No. 1), 65–75.
[7] Taken from Glaser, B. and Strauss, A. (1965) *Awareness of Dying* (Chicago: Aldine).
[8] Western, K. A. (Chap. 2, note 2). Western takes his definition from Beach, H. D. (1967) *Management of Human Behavior in Disaster* (Canada, Department of National Health and Welfare). A further recent definition of disaster, which is intended to orient discussions about the provision of relief for disasters, is provided by Skeet, who defines disaster in her manual as: 'An occurrence of such magnitude as to create a situation in which the normal patterns of life within a community are suddenly disrupted, and people are plunged into helplessness and suffering, and, as a result, may urgently require food, shelter, clothing, medical attention, protection, and other life-sustaining requirements' Skeet, M. (1977) *Manual for Disaster Relief Work* (London: Churchill Livingstone).
[9] Carr, L. J. (1932), Disaster and the sequence pattern concept of social change, *American Journal of Sociology*, **38**, 207–218.
[10] Killian, L. M. (1956) *An Introduction to Methodological Problems of Field Studies in Disaster*, Committee on Disaster Studies, Report No. 8, Publication 465. (Washington D.C.: National Academy of Sciences, National Research Council.)
[11] For example, Fritz, C. E. and Marks, E. S. (1954) The NORC studies of human behavior in disasters, *Journal of Social Issues*, **10** (No. 3), 26–41.
[12] Just how small a subdivision of society this definition might be extended to remains a moot point.
[13] Hale, A. R. and Hale, M. (1972) Accidents in perspective, *Occupational Psychology*, **46**, 69–78.
[14] This was one of the problems in the situation prior to Pearl Harbor discussed by Wohlstetter, R. (1962) *Pearl Harbor: Warning and Decision* (Stanford: Stanford University Press).
[15] See Wolfenstein, M. (1957) *Disaster: a psychological essay* (London, Routledge and Kegan Paul); and Lawrence, A. C. (1974) Human error as a cause of accidents in gold-mining, *Journal of Safety Research*, **6**, 78–88; esp. 87.
[16] Lawrence, A. C. (note 15), pp. 84–85.
[17] In considering parallels between disasters and detective novels, it is fascinating to note that B. Rouché comments in the introduction of his literary presentation of the events leading up to epidemics of various kinds:
'If the form of the narratives in *Annals of Epidemiology* seems to resemble that of the classic detective story, it should be remembered that Sir Arthur Conan Doyle derived the Holmesian method from that of the great Edinburgh diagnostician, Dr Joseph Bell'; Rouché, B. (1967) *Annals of Epidemiology* (Boston: Little, Brown) p. x.
Biographers of Doyle, however, have questioned whether Bell was as able as Doyle recalled. See Pearson, H. (1943) *Conan Doyle: his life and art* (London: Methuen).
Considering the transformations associated with accidents and disasters from

a slightly different point of view, we may draw attention to the similarities between the character of a precipitating incident set out above, and the 'triggers' which the sociologist Goffman discusses as provoking a switch from one 'frame' of understanding to another. See Goffman, E. (1975) *Frame Analysis* (London: Penguin). It is relevant also to later discussions to observe that Goffman developed his notions of frame analysis from Bateson's observations on a number of aspects of learning phenomena. See Bateson, G. (1973) *Steps to an Ecology of Mind: Collected Essays on Anthropology, Psychiatry and Epistemology* (St Albans: Granada). Another parallel which is further developed in later chapters is that offered by the transformations dealt with mathematically by 'catastrophe theory' although there is a need for considerable caution in approaching the translation of this mathematical treatment satisfactorily into the sociological realm. See Chapter 8. Note also the parallels between the effect of the precipitating incident and the properties of what have been referred to as 'summary events' by Rosengren *et al.* (Chap. 3, note 13) in their discussion of the Barsebäck panic.

[18] Lawrence, A. C. (Chap. 2, note 72), p. 84, Table 2.
[19] Aberfan report (see Appendix) paras. 109–111, 141, 146, 172.
[20] London Smallpox report (see Appendix) para. 180.
[21] Carr, L. J. (Chap. 3, note 1).
[22] Wolfenstein, M. (Chap. 2, note 1).
[23] See, for example, discussions in Wolfenstein, M. (Chap. 2, note 1), or in Barton, A. H. (1969) *Communities in Disaster: a sociological analysis of collective stress situations* (London: Ward Lock).
[24] Barton, A. H. (note 23) Chapter VI.
[25] Coldharbour Fire report (see Appendix).
[26] Cambrian Colliery report (see Appendix), para. 67.
[27] *Ibid.* para. 64. For a discussion of the difficulties of methane detection in this case, see paras. 71–82.
[28] *Ibid.* paras. 33, 37.
[29] *Ibid.* paras. 33, 37.
[30] *Ibid.* paras. 33–35.
[31] Lawrence, A. C. (Chap. 2, note 72), pp. 78–79.

Chapter 6

[1] At points in the research process like this, there is of course a considerable problem, for selective perception may well influence the handling of evidence in a way which reinforces prejudices and errors formed in the early stages of the work, but there seems to be no way of avoiding this hazard, apart from attempting to maintain some awareness of it, and even then, such awareness cannot hope to pick up all biases. This hazard is inherent in the method being used, and it therefore seems to be preferable to make it explicit. Although no solution is available to the single investigator, it is to be hoped that, in the long run, the collective judgement of others will weed out the worst misperceptions. Later commentary upon a misapprehension of De Tocqueville's about the American inheritance system is instructive in this regard. See Pierson, G. W. (1959) *Tocqueville in America* abridged by Lunt, D. C. (New York: Doubleday) esp. p.

85. The process of the development of such individual and collective misapprehensions and their more or less painful correction is central, of course, to the whole of the present approach to the study of disasters and unexpected disruptions, and this issue is developed in later chapters.

[2] Some of the material presented in the early sections of this chapter has been set out in Turner, B. A. (1976) The organizational and interorganizational development of disasters, *Administrative Science Quarterly*, **21** (No. 3), 378–397.

[3] This issue is developed further in the discussion of the revision of decision-premises in Chapter 9.

[4] See Benson, J. K. (1975) The interorganizational network as a political economy, *Administrative Science Quarterly*, **20** (No. 2), 229–249.

[5] See the discussion of threats and warnings in Chapter 3, above.

[6] Note that this general area is very similar to that associated with conditions of disjunct information, and could readily be classified under Category B in the classification set out earlier in this Chapter, so that the following discussion is, in part, an elaboration of that category.

[7] Six types of official report relating to non-military accidents and disasters are listed among British Government publications: reports into civil air accidents, boiler explosions, railway accidents, merchant shipping wrecks, mining accidents and other miscellaneous incidents discussed in reports presented as House of Commons papers or Command papers. In the present investigation, no attempt was made to examine the extensive set of reports on air accidents or the separate reports on fishing vessel wrecks, and only one rail accident report was analysed, but all of the reports in the remaining four classes for the period 1965–75 were analysed. For further details of the reports analysed, and the mode of analysis adopted, see Turner, B. A. (1976) *The Failure of Foresight* (Ph.D. thesis, University of Exeter).

[8] The account presented is drawn from the *Report of the Committee appointed to inquire into the circumstances, including the production, which led to the use of contaminated infusion fluids in the Devonport section of Plymouth General Hospital* (1972) Cmnd 4470 (London: HMSO).

[9] Cf. Turner, B. A. (March 1976) How to organize disaster: a new discipline brings organizational ruin within the grasp of every manager, *Management Today*, 56–57 and 105.

[10] The account presented is drawn from *Report of the Committee of Inquiry into the Smallpox Outbreak in London in March/April 1973* (1974) Cmnd 5626 (London: HMSO).

[11] The account presented is drawn from the report: *Public Inquiry into a fire at Dudgeon's Wharf, on 17 July 1969* (1970) Cmnd 4470 (London: HMSO).

[12] The account presented is drawn from *Marine Wreck Report No. 8052 m.v. Nicolaw* (1971) Department of Trade and Industry (London: HMSO).

[13] The account presented is drawn from *Marine Wreck Report No. 8054 m.t. Esso Ipswich and m.t. Efthycosta II* (1971) Department of Trade and Industry (London: HMSO).

[14] The account presented is drawn from *Marine Wreck Report No. 8059, m.v. Redthorn and m.v. Efpha* (1973) Department of Trade and Industry (London: HMSO).

[15] The account presented is drawn from *Marine Wreck Report No. 8063. m.v.*

British Fern and m.v. Teviot (1975) Department of Trade and Industry (London: HMSO).

[16] Of McGregor, an able seaman on the deck of the *Teviot* who could have been told to keep a lookout, the Court commented that he was, at the material time, 'on the poop deck, retrieving a wad he had dropped there. We found his oral evidence of little value. He would be a good raconteur in any gathering.'!

[17] It is interesting to place such events on a kind of continuum between situations, on the one hand, in which man is facing Nature as an adversary, so to speak; and war and conflict situations on the other hand, where there is an overt, two-party intent to cause harm to the other party. Setting aside the evident differences, it can be seen that there are similarities in the manner in which the actions of the other party are likely to be scrutinized to divine intentions and outcomes, and in the manner in which feints, ambiguities, decoys, concealment of evidence and other means of misleading the other party come to be of crucial importance. See Scott's discussion of recording, control, 'uncovering' and 'recovering' moves in information games in a horse-racing context: Scott, M. B. (1968) *The Racing Game* (Chicago, Aldine). For a discussion of more overt conflicts, see A. Rapoport's informative (1968) Introduction to *Clausewitz on War*, ed. Rapoport, A. (Harmondsworth: Penguin) pp. 11–82.

[18] See, for example, the discussion of boundaries in Miller, E. J. and Rice, A. K. (1967) *Systems of Organization* (London: Tavistock), Chapter 2.

[19] For an extended discussion of organizational subcultures, see Turner, B. A. (1971) *Exploring the Industrial Subculture* (London: Macmillan).

[20] See the discussion of the approaches to the study of rationality made by Simon and others in Chapter 7, below.

[21] Burns, T. and Stalker, G. M. (1966) *The Management of Innovation*, Second edition (London, Tavistock) p. 155. See also Turner, B. A. *Exploring the Industrial Subculture* (note 19), pp. 38 and 160.

[22] An approach to the analysis of organizations based upon the task which they carry out is discussed in *Industrial Organization: behaviour and control* (1970) Woodward, J. ed. (Oxford: Oxford University Press).

[23] As in Japanese organizations, for example: Abegglen, J. C. (1958) *The Japanese Factory* (New York: Free Press).

[24] As in dock work, for example: *The Dockworker* (1954) ed. Woodward, J. (Liverpool: University of Liverpool Press).

[25] As in the organization studied by Dalton: Dalton, M. (1959) *Men Who Manage* (New York: Wiley).

[26] See Barnard, C. I. (1938) *The Functions of the Executive* (Cambridge, Mass: Harvard University Press).

Chapter 7

[1] For example Brillouin, L. (1949) Life, thermodynamics and cybernetics, *American Scientist*, **37,** 554–568; Brillouin, L. (1964) *Scientific Uncertainty and Information* (New York, Academic Press); Schrödinger, E. (1944) *What is Life?* (Cambridge, Cambridge University Press). For a discussion of some of the ways in which time in human experience may be regarded, see Cottle, T. J. and Klineberg, S. L. (1974) *The Present of Things Future: explorations of time in*

human experience (New York: Free Press) esp. pp. 3–35.
[2] Schrödinger, E. (note 1), p. 74.
[3] See Yockey, H. P., Platzmann, R. L. and Quastler, H. eds. (1958) *Symposium on Information Theory in Biology* (London: Pergamon).
[4] See, for example, Rorive, A. (1960) Théorie de l'information et structure des sociétés humaines, *Proceedings 2nd International Congress on Cybernetics, Namur, 1958* (Association Internationale de Cybernétique) pp. 611ff. See also Rothstein, J. (1958) *Communication, Organization and Science* (Indian Hills, Colorado: Falcons Wing Press).
[5] This distinction, which seems to be similar to the current biological distinction between 'teleology' and 'teleonomy' is made by Emmett, D. (1958) *Function, Purpose and Power: Some Comments on the Status of Individuals in Societies* (London: Macmillan) pp. 68–69.
[6] Compare Erikson, E. (1960) *Insight and Responsibility: lectures on the ethical implications of psychoanalytical insight* (London: Faber). See also Cottle, T. J. and Klineberg, S. L. (note 1).
[7] To use Simon's term: Simon, H. A. (1957) *Administrative Behavior*, 2nd ed. (New York: Free Press).
[8] The range of such models is suggested in an absorbing article by the self-styled 'ludiologist', Miller, S. N. (1974) The playful, the crazy and the nature of pretence, *The Anthropological Study of Play, Rice University Studies*, **60** (No. 3), 31–51.
[9] See, for example, the discussion by Luckman, T. (1975) On the rationality of institutions in modern life, *European Journal of Sociology*, **XVI** (No. 1), 3–15.
[10] For some discussions of problems associated with the concept, see, for example, Schutz, A. (1964) The problem of rationality in the social world, *Studies in Social Theory* Vol. II *Collected Papers* ed. Brodersen, A. (The Hague: Nijhoff) pp. 64–88; and Wilson, R. (1970) *Rationality* (Oxford: Blackwell).
[11] See Parsons, T. (1937) *The Structure of Social Action* 2 vols. (New York: Free Press). Also Turner, B. A. (1971) *Exploring the Industrial Subculture* (London, Macmillan) pp. 35–38.
[12] See pp. 133–6 below.
[13] See, for example, Mannheim, K. (1940) *Man and Society in an Age of Reconstruction: studies in modern social structure* (London: Kegan Paul, Trench, Trubner).
[14] *Ibid.* p. 53.
[15] *Ibid.* p. 55.
[16] *Ibid.* p. 52.
[17] *Ibid.* p. 58.
[18] *Ibid.* p. 147.
[19] See, for example, the discussion in Merton, R. K. (1961) Epilogue, *Social Problems and Sociological Theory* eds. Merton, R. K. and Nisbet, R. A. (New York: Harcourt), pp. 697–737, esp. footnote, p. 711.
[20] Compare the comment by Henderson, L. J. 'For the present, we are as a rule unable to foresee even the more important results of an experimental modification of the social system, and ... there is no prospect that we shall presently be able to foresee them'. Henderson, L. J. (1970) Sociology 23 lectures, *L. J. Henderson on the Social System: selected writings* (Chicago: University of

Chicago Press) ed. Barber, B. pp. 57–148, 146.

[21] This point is made by Merton, R. K. in his classic paper (1936) The unanticipated consequences of purposive social action, *American Sociological Review*, **1** (No. 6), 894–904. Some of the points made below are taken from Merton's discussion.

[22] See Selznick, P. (1952) A theory of organizational commitments, *Reader in Bureaucracy* ed. Merton, R. K. *et al.* (New York, Free Press), pp. 194–202.

[23] Merton, R. K. (note 21).

[24] Selznick, P. (1966) *TVA and the Grass Roots: a study in the sociology of formal organizations* (Berkeley: University of California Press).

[25] For a fuller discussion of these issues, see Chapter 6.

[26] There is a degree of uncertainty about the point at which this distinction first emerges in Simon's work, Storing and others pointing to a shift in Simon's position in the second edition of his *Administrative Behavior*, while Simon maintains that the distinction was implicit in the first edition. A similar ambiguity surrounds the question of whether Simon was originally concerned with 'decisions' or with 'decision premises'. See Storing, H. J. (1962) The science of administration: Herbert A. Simon, *Essays on the Scientific Study of Politics*, ed. Storing H. J. (New York: Holt, Rinehart and Winston) pp. 63–150; and, for a more recent discussion from a standpoint informed by ethnomethodology, Wilson, H. T. (1973) Rationality and decision in administrative science, *Canadian Journal of Political Science*, **VII** (No. 2), 271–294. The discussion in this chapter and in later chapters starts from Simon's later position, and goes some way towards compensating for the individualistic assumptions which Simon holds although not as far, presumably, as Wilson would like.

[27] Simon, H. A. (1957) *Models of Man, Social and Rational: mathematical essays on rational human behavior in social settings* (New York: Wiley), p. 198. Again, there is some dispute about the extent to which Simon's Principle is implicit in his earlier work, or is a later addition made necessary by the initial inadequacies of his approach. See again Storing, H. J. (note 26), and Wilson, H. T. (note 26). Whatever the origins of this principle, it states very clearly an important issue in the present discussion.

[28] Wilson (note 26) suggests that the two forms of rationality discussed result from Simon having to abandon a commitment to what he calls, in a phrase taken from Simon himself, 'a preposterously omniscient rationality', p. 277.

[29] In the 'Introduction' (1957) to the second edition of *Administrative Behavior*: Simon, H. A. *Administrative Behavior* 2nd edn. (New York: Free Press).

[30] As Wilson (note 26) again points out, p. 276.

[31] A strategy of gradually increasing or extending present practices or solutions rather than starting again from the beginning. For discussions of incrementalism in the context of planning theory, see Friedman, J. and Hudson, B. (1974) Knowledge and action: a guide to planning theory, *Journal of the American Institute of Planning*, **40** (No. 1), 2–16; and Berry, D. (1974) The transfer of planning theories to health planning practice, *Policy Sciences*, **5**, 343–361.

[32] Simon's approach has been criticized as conjuring up a picture of an 'eternal escalator' of rationalities of means which 'ascends and ascends but arrives nowhere', because every end turns out to be a means to something beyond it. See

Waldo, D. (June 1952) Replies and comments, *American Political Science Review*, **46**, 503. However, it seems to be possible to adopt some aspects of his hierarchical approach to rationalities and to situate them in organizational, institutionalized settings without taking over at the same time the assumption that all ends are given for the administrator.

[33] Cf. Selznick, P. (note 22), footnote p. 198.

[34] Selznick has pointed to the interrelationship between individual and collective ignorance: 'The tendency to ignore factors not considered by the formal system', he notes, 'is inherent in the necessities of action and can never be eliminated'. Selznick, P. (note 22), p. 198 footnote. On the related issue of the inability of any complex system to produce a full description of itself, see Bateson, G. (1973) Style, grace and information in primitive art, in *Steps to an Ecology of Mind: collected essays on anthropology, psychiatry, evolution and epistemology* (St Albans: Granada) pp. 101–125.

Chapter 8

[1] Following the work of Shannon, C. E. and Weaver, W. (1949) *The Mathematical Theory of Communication* (Urbana, Ill: University of Illinois Press).

[2] The most common mathematical expression for 'Shannonian' information (H) is:

$$H = -p(i) \log_2 p(i)$$

and this measure, with or without the sign reversed, is used also to measure certainty or uncertainty. These measures are essentially the same since they are measures of the number of binary decisions needed to assign a given message to a specified number of empty categories, given that the receiver knows the probabilities of the items being transmitted in each of these categories. The point has been made by Cronbach that there are other measures of uncertainty or information, and the user of Shannonian information has it incumbent upon him to demonstrate, if he is seriously using this as a scientific measure, that he is concerned with message space, rather than with certainty as more colloquially understood (since *log* uncertainty is a measure of message space in 'bits'); that the receiver knows the probabilities and joint probabilities of the source; that infinitely long messages are being considered, and infinite delay is allowed; and all categories and errors are equally important: Cronbach, L. J. (1955) On the non-rational application of information measures in psychology, in *Information Theory in Psychology: problems and methods* (Glencoe, Ill: Free Press) ed. Quastler, H. pp. 14–30. These assumptions are not being made here, and the Shannonian measure of information would not necessarily apply. What is assumed is the relationship that equates the amount of information transmitted with a gain in certainty, or a reduction in uncertainty, a relationship which, within limits, should presumably hold true, whatever measures of information and certainty are used. See Quastler, H. 'Information theory terms and their psychological correlates' *ibid.* pp. 143–173. For discussions of some of the limitations of Shannonian information, see the (1977) articles collected in Issue No. 1 *Cybernetica*, **XX**.

[3] See the discussion by Cherry, C. (1957) *On Human Communication: a*

review, a survey and a criticism (London: Chapman and Hall) esp. p. 216. For other general discussions of information theory, see Quastler, H. (1958) A primer on information theory, in *Symposium on Information Theory in Biology* ed. Yockey, H. P., Platzman, R. L. and Quastler, H. (London: Pergamon) pp. 3–49; Quastler, H. (1955) (see note 1); and Edwards, E. (1964) *Information Transmission: an introductory guide to the application of the theory of information to the human sciences* (London: Chapman and Hall).

[4] See Quastler, H. (note 2).

[5] Although, as Simon has rightly pointed out, it is unreal to expect individuals to have fully formed probability distributions in their heads as guides for action.

[6] Although, as Cronbach (note 1) has pointed out, it might not be appropriate in such a case to use the Shannonian expression to measure information of this form. This point does not materially affect the argument here. See Note 2, above.

[7] By Bakan, who says that 'not all that can be quantified is automatically tautologous' Bakan, D. (1974) 'Mind, matter and the separate reality of information', Presidential address to the American Psychological Association, 1972. *Philosophy of the Social Sciences*, **4**, 1–15.

[8] See, for example, Cherry, C. (note 3), p. 216.

[9] Aspects of the confusion surrounding the use of this term have been discussed by both Cherry, C. (note 3) and Cronbach, L. J. (note 1).

[10] See, again, Cherry, C. (note 3), who directs the reader at this point (p. 216) to work of MacKay, Gabor and Brillouin, but none of these writers, in his examination of the relationships between information and scientific process, is addressing the issue under discussion here. It is relevant, here, too, to note Churchman's comments on the assessment of probability. He states the impossibility of obtaining an objective definition of probability, with the consequence that judgement must always be reintroduced. MacKay, D. M. (1950) Quantal aspects of scientific information, *Philosophical Magazine*, **41**, 289ff.; Gabor, D. (1950) Communication theory and physics, *Philosophical Magazine*, **41** 1161ff.; Brillouin, L. (1951) Information theory and entropy I, *Journal of Applied Physics*, **22** (no. 3), 334ff. and Information theory and entropy II, *Journal of Applied Physics*, **22** (No. 3), 338ff. Churchman, C. W. (1961) *Prediction and Optimal Decision-making: philosophical issues of a science of values* (Englewood Cliffs, N.J. Prentice Hall) Chapter 6.

[11] Churchman, C. W. (note 10), p. 108.

[12] Unless a small range of unlikely events which Revere's men might readily acknowledge as feasible is to be included, such as the enemy retreating, the weather worsening so as to end the engagement, and so on. These rare events do not alter the measurement of future uncertainty, however, as Quastler has noted (Quastler, H., 1958; note 2). The popular account of Paul Revere's ride to Lexington in April 1775 is to be found in the poem written in 1863 by Longfellow: 'The Landlord's Tale—Paul Revere's Ride', Longfellow, H. W. *Poetical Works* (London: Warne, Chandos Classics edition, n.d.) pp. 363–366. I am grateful to Dr M. Gidley for this reference.

[13] See the references in Notes 2 and 3, above.

[14] Bakan, D. (note 7).

[15] Bakan, D. (note 7) refers to Laplace, Helmholtz and Darwin in this context, although Maruyama argues that Darwin held a more complex and subtle

view: Maruyama, M. (1974) Paradigmatology and its application to cross-disciplinary, cross-professional and cross-cultural communication, *Cybernetica*, **17** (No. 2), 136–156.

[16] See Eiseley, L. (1973) *The Unexpected Universe* (Harmondsworth: Penguin).

[17] See Bakan, D. (note 7).

[18] The communication theory model of information transmission, appropriately for the purposes for which it was developed, is concerned with mechanical, non-living processes, in which it is not even necessary to make any assumptions about the occurrence of feedback. Maruyama characterizes this approach as one typical of a classificational epistemology, which regards the universe as being made up of disjoined categories and sub-categories, an epistemology which Maruyama wishes to challenge. Maruyama (note 15); Maruyama (1974) Symbiotization of cultural heterogeneity: scientific epistemological and esthetic bases, *Co-existence*, **11**, 42–56. For other related challenges to the prevailing epistemology, see Pirsig, R. M. (1974) *Zen and the Art of Motorcycle Maintenance* (London: Bodley Head) and Witkin, R. W. (1974) *The Intelligence of Feeling* (London: Heinemann).

There is no consideration in Shannonian information theory of the active part played by the receiver of information with regard to the messages being relayed. As Maruyama puts it:

"In Shannon's theory, the amount of information is related to the degree of specificity, and is measured by the degree of improbability of the structure of the 'message' of the 'data' as contrasted to the combination of independent random events in the given context. If left to the influence of independent random events, structures will decay with a great probability. Independent random events are not likely to generate structures *steadily* and *systematically*. Therefore, in Shannon's formulation, information can only decrease. Evolution and growth of complexity are impossible in this formulation, or at least so highly improbable that their common occurrence has to be attributed to some processes which are inexplicable by Shannon's theory". Observe, however, that Maruyama here neglects Cherry's distinction between channels of communication and channels of observation.

[19] Szilard, cited by Bakan, D. (note 7).

[20] Kuhn, T. S. (1970) *The Structure of Scientific Revolution* (Chicago: Chicago University Press). For a discussion of the contribution of L. Fleck to the ideas presented by Kuhn, see Baldamus, W. (1976) *The Structure of Sociological Inference* (London: Martin Robertson) Chapter 1.3 and *passim*.

[21] See Hewish, A. (Oct. 1968) Pulsars, *Scientific American*, **219** (No. 4), 25–35; Wade, N. (Aug. 1975) Discovery of pulsars: a graduate student's story, *Science*, **189**, 358–364. In view of our discussions of surprise and precipitating incidents, it is interesting to note the way in which Jocelyn Burnell (née Bell) describes her recognition of the 'bit of scruff' on the record as something worthy of further investigation:

> When it clicked that I had seen it before, I did a double click. I remembered that I had seen it for the same part of the sky before. This bit of scruff was something I didn't completely understand—my brain just hung on to it and I remembered that I had seen it before. *Ibid.* p. 359.

Dean has reviewed this case as an example of serendipitous discovery, in relation to Kuhn's work, but he uses the term 'serendipitous' to denote discoveries that are 'unanticipated *and* yet prove compatible with existing scientific assumptions' p. 83, footnote 1 (Dean's emphasis). This is slightly different from the usage adopted here, in that we are restricting the term to those discoveries which do provoke a reassessment of existing theories. The discrepancy in usage is not crucial, however, for the point which Dean is pursuing, of whether such discoveries form part of 'normal' science or not is parallel to the issue of whether information is surprising or not. While in both cases we can see different levels of discovery and surprise, there seems to be no satisfactory way of drawing a line between 'normal' and 'non-normal' science, or between slightly surprising and very surprising information, in a rigorous way that will cover all borderline cases. See Dean, C. (1977) Are serendipitous discoveries a part of normal science? The case of the pulsars, *Sociological Review*, **25** (No. 1), 73–86.

[22] Hixon report para. 193.

[23] In the way that incorrect responses associated with Binet's early intelligence tests provided a fruitful area of inquiry for the young Piaget. Flavell, J. H. (1963) *The Developmental Psychology of Jean Piaget* (London: Van Nostrand), p. 3.

[24] But see the comments already made in note 21, above.

[25] Some examples of surprise relevant to accidents are discussed in the following chapter.

[26] Anticipating the terms used in Chapter 9, we could relate the unexpectedness to the extent of the revision of 'bounded decision zones' provoked.

[27] See, for example, the confrontations and lobbying over issues relating to the development and extension of nuclear power represented by publications such as the Report of the Royal Commission on Environmental Pollution (1976) *Nuclear Power and the Environment* (London: HMSO); Patterson, W. (1976) *Nuclear Power* (Harmondsworth: Penguin); Friends of the Earth (1976) *Nuclear Prospects: a comment on the individual, the state and nuclear power* (London: Friends of the Earth). Recent pressures brought to bear by scientists for a relaxation of the new rules pertaining to the control of genetic manipulation in Britain and in the United States offer an instance of lobbying moving in the opposite direction.

[28] A simple mathematical model for coping with some aspects of differential surprise is presented in Turner, B. A. (1976) *The Failure of Foresight* (Ph.D. thesis, University of Exeter) Annex to Chapter IX.

[29] Thom, R. (1975), *Structural Stability and Morphogenesis: an outline of a general theory of models*. Translated by Fowler, D. H. (London: W. A. Benjamin). See also Zeeman, E. C. (1976) Catastrophe theory. *Scientific American*, **234** (No. 4), 65–83; and Stewart, I. (20 Nov. 1975). The seven elementary catastrophes, *New Scientist*, **68**, 447–454.

[30] Cf. Professor Waddington's comment in the Foreword to Thom's book: 'I cannot claim to understand all of it; I think that only a relatively few expert topologists will be able to follow all his mathematical details...' p. xix in Thom, R. (note 29).

[31] Kuhn, T. S. (note 20). See also the development of Kuhn's ideas in relation to aspects of catastrophe theory in Turner, B. A. *The Failure of Foresight* (note 28), pp. X.21–X.26.

[32] See Bateson, G. The logical categories of learning and communication in Bateson, G. (1973) *Steps to an Ecology of Mind: collected essays in anthropology, psychiatry, evolution and epistemology* (St Albans: Granada), pp. 250–279.
[33] Goffman, E. (1975), *Frame Analysis* (Harmondsworth: Penguin).
[34] See, for example, discussions by Garfinkel, H. (1967) *Studies in Ethnomethodology* (Englewood Cliffs, N.J.: Prentice Hall); and by McHugh, P. (1968) *Defining the Situation: the organization of meaning in social interaction* (Indianapolis: Bobbs–Merrill). Similar psychological processes have also been explored by Piaget, who talks of vertical and horizontal *'decalages'*, the vertical ones being of particular relevance here. See Flavell, J. H. (note 23), pp. 20–23 and 58. For an alternative approach to psychological discontinuities, see Isaac, D. J. and O'Connor, B. M. (1975) A discontinuity theory of psychological development, *Human Relations*, **29** (No. 1), 41–61.
[35] Platt, J. (Nov. 1970), Hierarchical growth, *Bulletin of the Atomic Scientists*, 2–4 and 46–48.
[36] This suggestion comes from Deutsch: Deutsch, K. W. (1966) *The Nerves of Government: models of political communication and control* (New York: Free Press). One of the outstanding issues to be followed up is the question of just how the major sub-system around which organization is focused can be specified, if Deutsch's suggestion is correct. Giddens makes a similar point when he refers to invariances across paradigms, using the example of essential similarities which link Roman Catholicism with Protestantism, in spite of the schism which separates them. Giddens, A. (1976) *New Rules of Sociological Method: a positive critique of interpretative sociology* (London: Hutchinson).

Chapter 9

[1] In this discussion, the possible consequences of the behaviour of crowds is being set aside. Large-scale accidents do occur from time to time as a result of the collective actions of the individual members of a crowd, leading to incidents such as the Bethnal Green disaster in 1943, and the Ibrox Park disaster in 1971. Smelser, in the formulation of his 'value-added' model of collective behaviour, has made a valuable contribution to the understanding of the manner in which collective behaviour in crowds and related groupings may give rise to undesirable outcomes, and his analysis seems to be complementary to the one presented here. See Smelser, N. J. (1962) *Theory of Collective Behaviour* (London: Routledge and Kegan Paul); also *Report of an Inquiry into the Accident at Bethnal Green Tube Station Shelter on 3 March 1943* (1943) Cmnd 6583 (London: HMSO).
[2] For an introduction to this kind of approach, see, for example, Luce, D. and Raiffa, H. (1957) *Games and Decisions* (New York: Wiley); Duncan, W. J. (1973) *Decision-making and Social Issues: a guide to administrative action in an administrative context* (Hinsdale, Ill: Dryden Press); Cesario, F. J. (1975) A primer on entropy modelling, *Journal of the American Institute of Planners*, **41** (No. 1), 40–48.
[3] Compare the attempts by Halliday to estimate the level of hazard likely to arise from experiments with 'genetic manipulation' and the comment made in

subsequent correspondence: Halliday, R. (17 Feb. 1977), Should genetic engineers be contained?, *New Scientist,* **73** (No. 1039), 399–401; and letter by Urban, M. R. (3 March 1977), Genetic engineering, *New Scientist,* **73** (No. 1041), 550–551.

[4] For an enlightening discussion of some of these difficulties, together with an account of some relevant psychological experimental evidence, see Fischoff, B. (1975) Hindsight: thinking backward?, *Psychology Today,* **8**. The discussion by Dingwall of Sacks' ethnomethodological consideration of the 'artful nature' of *post hoc* experience indicates an approach to the same phenomenon from a different direction. See Dingwall, R. (1976) *Aspects of Illness* (London, Martin Robertson) Chapter 4, pp. 61–95.

[5] Compare the discussion by Owen of the way in which men, prior to the Houghton Main Colliery disaster could behave 'reasonably' in switching on a fan, even in circumstances where to do so might constitute a breach of regulations. See Owen, D. (1977) Some Organizational Aspects of a Public Inquiry: Working Paper, Mimeo. Department of Management Studies (Coventry: Lanchester Polytechnic).

[6] Although Simon addresses himself explicitly to the problems faced by 'administrative man', it is necessary to fill out some aspects of his analysis which he himself does little more than hint at, particularly with regard to the way in which constraints upon the individual are related to constraints upon the organization. Simon does not neglect these problems, but many of his discussions refer to the organization of behaviour by 'information processing organisms' at a general level which implies that his comments will apply equally to animals, men, automata, computers and to organizations. Although we are here more specifically concerned with the behaviour of men in organizations, the following analysis does, nonetheless, lean heavily upon Simon's work.

[7] Simon suggests the kind of link between the individual decision-maker and the organization which is being referred to here in a crucial comment: he proposes that for individuals adopting roles within organizations that are subject to the constraints of rationality, 'the specification of a role consists in the specification of some subset of the premises that are to guide the decisions of the actor as to his course of behaviour'. Simon, H. (1957) *Models of Man* (New York: Wiley), p. 201. As the previous note suggested, much of the present chapter may be regarded as an elaboration of this comment, which Simon himself does not develop. See also the discussion of the negotiation of roles in organizations in Turner, B. A. (1971) *Exploring the Industrial Subculture* (London: Macmillan), pp. 69–91.

[8] A witty guide to the rhetoric of academic decision-making is presented in the little classic: Cornford, F. M. (1908) *Microcosmographica Academica: being a guide for the young academic politician* (Cambridge: Bowes and Bowes).

[9] See Turner, B. A. *Exploring the Industrial Subculture* (note 7).

[10] This approach towards conflict resolution is taken by Lawrence, P. R. and Lorsch, J. W. (1967) *Organization and Environment: managing differentiation and integration* (Cambridge, Mass: Harvard Graduate School of Business Administration).

[11] See Turner, B. A., *Exploring the Industrial Subculture* (note 7).

[12] The phrase is taken from Berger, P. and Luckmann, T. (1967) *The Social Construction of Reality* (London: Allen Lane).

[13] See Lawrence, P. R. and Lorsch, J. W. *Organization and Environment* (note 10).
[14] There are parallels between the above notion of a bounded decision zone and Abell's recently developed idea of 'influence zones'. It is possible that some of Abell's mathematical models could be adapted to cope with the present concept, if a mathematical treatment were desired. I am indebted to Professor D. Hickson for drawing my attention to this possibility. See Abell, P. ed. (1975) *Organizations as Bargaining and Influence Systems* (London: Heinemann) esp. the earlier chapters by Abell.
[15] Different types of institution may vary in the extent to which they hold rigid institutionally approved views of the world, and in the flexibility of their response to new events. Thus, it may well be difficult completely to 'surprise' a research team with new events within the field which they are studying, for although, inevitably, they will have their collective *idées fixes*, they are likely also to have available to them a series of provisional views of the world, a number of possible hypotheses which they could credibly entertain at an institutional level, so that an event would have to be very unexpected indeed if they were to be caught without some framework which they could readily adopt to cope with it. But this is because one of the collective tasks tackled by research teams is the pursuit and examination of new and more appropriate frameworks of understanding. The situation of organizations which are less research-minded seems to be rather different, for those institutions which are committed to the attainment of more immediate, practical goals in the world are likely to feel that they cannot afford to retain such a high degree of flexibility in their institutional outlook. If they are to get things done, they have to become more single-minded, and to simplify their perception of the world to this end. Inevitably, as a result of this commitment, therefore, they become more vulnerable to surprise.
[16] Note that the organization is here being regarded as an entity with perfect information flows internally. A similar treatment could, however, be reworked for a department or for a workgroup, with the rest of the organization treated as part of the group's environment.
[17] The account presented is drawn from *Marine Wreck Report No. 8043 m.v. Anzio I* Department of Trade and Industry (1967) (London: HMSO).
[18] The account presented is drawn from *Marine Wreck Report No. 8033 m.v. Starcrown* Department of Trade and Industry (1965) (London: HMSO).
[19] The account presented is drawn from *Marine Wreck Report No. 8045 m.v. Isle of Gigha*, Department of Trade and Industry (1967) (London: HMSO).
[20] The account presented is drawn from *Accident at Markham Colliery, Derbyshire (30 July 1973)* Command paper (1974) Cmnd 5557 (London: HMSO).
[21] The account presented is drawn from the *Report of the Inquiry into the collapse of flats at Ronan Point, Canning Town*, Ministry of Housing and Local Government (1968) (London: HMSO).
[22] For a discussion of some of the problems of rationality in relation to accidents, see Eldridge, J. and Kaye, B. M. (1973) Wages and accidents: an exploratory paper, in *Onderneming en vakbeweging, Mens en Maatschappij*, ed. Teulings, A. (The Hague) pp. 151–171.
[23] This is very much a working assumption, of course, intended to exclude

from the present study the need to examine this whole set of issues. For a report of some of the considerations raised by an examination of the 'reasonableness' or otherwise of public inquiries, see Owen 'Some Organizational Aspects of a Public Inquiry: Working Paper' (note 5). The research reported by Owen is an examination of the processes of public inquiries, which scrutinizes one particular case in an intensive manner.

[24] Assuming, of course, that the option of not permitting ships to put to sea at all is ruled out.

[25] Further revisions have subsequently been suggested in a second report (1976): *Safety Manriding in Mines: First Report of the National Committee for Safety of Manriding in Shafts and Unwalkable Outlets* (London: HMSO).

[26] Or so a press interview with one of the victims of the Cocoanut Grove fire reported.

[27] *Ronan Point Inquiry Report* (see Appendix) para. 14.

[28] The term 'plan' is taken from the usage of Miller, Galanter and Pribram who, significantly, developed their psychological approach to planful behaviour from earlier work by Simon and his colleagues. At this point, it might be appropriate to note that while the approach taken by Miller *et al.* has been described as a landmark in the evolution of rationalist thought (by Friedmann and Hudson), it has been criticized by Maruyama for failing to discuss what Maruyama calls 'deviation-amplifying feedback', so that the approach has difficulty in dealing with creativity and innovation, topics which are of some significance to the present discussion. See Miller, G. A. Galanter, E. and Pribram, K. H. (1960) *Plans and the Structure of Behavior* (New York: Holt, Rinehart and Winston); Friedmann, J. and Hudson, B. (1974) Knowledge and action: a guide to planning theory, *Journal of the American Institute of Planners*, **40** (No. 1), 2–16; Maruyama, M. (1963) The second cybernetics: deviation amplifying mutual causal processes, *American Scientist*, **51,** 164–179.

[29] Explicit parallels are being drawn here between the analysis of the consequences of errors arising within organizational hierarchies and recent work carried out by psychologists investigating the failure of human intent as evidenced by typing errors and speech errors; Fromkin, V. (1973) Slips of the tongue, *Scientific American*, **229** (No. 6), 110–117; Fromkin, V. (1973) *Speech Errors as Linguistic Evidence* (The Hague: Mouton); MacKay, D. G. (1970) Spoonerisms: the structure of errors in the serial order of speech, *Neuropsychologia*, **8,** 323–350; MacKay, D. G. (1972) The structure of words and syllables: evidence from errors in speech, *Cognitive Psychology,* **3,** 210–227; Shaffer, H. (1975) Control processes in typing, *Quarterly Journal of Experimental Psychology*, **27,** 419–432.

This approach has also been tentatively extended to an examination of the problems of skill learning and of accidents caused by failures in such learning: see Reason, J. T. (1977) Skill and error in everyday life, in *Adult Learning: psychological research and applications,* ed. Howe, M. J. A. (London: Wiley), Ch. 2. Following Lashley, K. S. it is assumed that speech, typing and other responses are produced by means of hierarchically organized mental and physiological processes. Errors may thus be regarded as non-random outcomes which can be related to the point at which the 'plan' of intent was 'corrupted', using the term 'plan' here again in the sense suggested by Miller *et al.* See Lashley, K. S. (1951)

The problem of serial order in behaviour, in *Cerebral Mechanisms in Behavior*, ed. Jeffress, L. A. (New York: Wiley); and Miller *et al.*, *Plans and the Structure of Behavior* (note 28).

[30] On physiological and psychological errors, see the comments in note 29.

[31] See again the sources mentioned in note 29. The suggested principle is adapted from Shaffer's comment that 'Speech errors make non-random use of the rules of language' Shaffer, H. Tips of the slung (January 1976) Public lecture given at Exeter University.

[32] The turning of the organizing propensity upon the organizer is an essential element in all of the Frankenstein/sorcerer's apprentice genre of stories and myths.

[33] This somewhat ironic proposition is pursued in Turner, B. A. (March 1976) How to organize disaster: a new discipline brings organizational ruin within the grasp of every manager, *Management Today*, 56–57, 105.

[34] We may compare this point directly with Thom's comment that biological mutation is not a purely random event, and that, out of two possible mutations, the mutation that minimizes the production of entropy or disorder is the most likely to occur. Thom, R. (1975) *Structural Stability and Morphogenesis: an outline of a general theory of models*, Trans. Fowler, D. H. (London: W. A. Benjamin), p. 282.

[35] BBC television news reports, June 1977.

[36] Even in the production of poetry and dreams: Eiseley talks of living matter as 'the cell that had somehow mastered the secret of controlled energy, of surreptitious burning to a purpose' (p. 176); and of life as the taking over of the oxidation process, controlled burning to produce dreams and poetry (p. 180). Eiseley, L. (1973) *The Unexpected Universe* (Harmondsworth: Penguin).

[37] For a useful general discussion of man's activities in harnessing energy, and of the problems that these activities give rise to, see Foley, G. (with Nassim, C.) (1976) *The Energy Question* (Harmondsworth: Penguin).

[38] See Haddon, W. Jr. (1968) The changing approach to the epidemiology, prevention and amelioration of trauma: the transition to approaches etiologically rather than descriptively based, *American Journal of Public Health*, **58** (No. 8), 1431–1438; and Haddon, W. Jr. (1973) Energy damage and the ten countermeasure strategies, *Human Factors*, **15** (No. 4), 355–366.

[39] Haddon's work has been criticized as being based upon an over-simple premiss, in taking energy alone to be the source of accidents. This, it has been commented, leads one to conclude that the injury due to falling against a piece of glass is due to the energy of one's own body, and not to the fact of the glass being exposed in the first place. See Martin, J. (1976), *Engineering Reliability Techniques* (Milton Keynes: Open University Press), p. 17. This criticism seems to be somewhat misplaced, for if the piece of glass is stationary, the source of the energy producing the injurious transformation must be in the body which falls against it. Rather than criticizing Haddon for including energy, therefore, it seems to be more appropriate to point to his crucial omission, as we are doing in this chapter: he does not consider the information or lack of information available about the circumstances in which the potential energy might be discharged.

[40] The close link between information and energy was pointed to in Maxwell's early discussion of the Second Law of Thermodynamics, when the energy

needed by his hypothetical 'demons' to gather and act upon information about the speeds of gas molecules, if they were to partition a container-full of medium-temperature gas into 'hot' molecules and 'cool' molecules, served to illustrate the requirements of the Law. See, for example Jeans, Sir J. (1933) *The New Background of Science* (Cambridge: Cambridge University Press), pp. 262–298; and Brillouin, L. (1964) *Scientific Uncertainty and Information* (New York: Academic Press) Ch. 13. Since then the information aspect has been developed by a number of writers, notably Brillouin, following upon Shannon's formulation of a mathematical measure of information content. The links between energy, information and order are particularly crucial for those studying living matter, for it is central to the processes by which living organisms turn energy in the world to their own advantage, to build up their own internal and external order. Information thus has an 'ontic status', as we noted in Chapter 8. See Bakan (1974) Mind, matter and the separate reality of information, Presidential address to the American Psychological Association, 1972: *Philosophy of the Social Sciences*, **4**, 1–15.

Chapter 10

[1] Some of the material in this chapter has already appeared, in a slightly modified form, as The origins of disaster in *Safety at Work: Recent Research into the Causes and Prevention of Industrial Accidents* (1977) Centre for Socio-Legal Studies, Wolfson College, Oxford, Conference Papers No. 1, ed. Philips, J. (Oxford: Social Science Research Council), pp. 1–18.
[2] This classification is derived from that presented by Western K. A. (Chap. 2, note 2).
[3] Bakan appears to have been responsible for coining the appallingly appealing jargon term 'negentropophagic' to describe 'order-eating beings'. Bakan, D. (1974) Mind, matter and the separate reality of information Presidential address to the American Psychological Association, 1972: *Philosophy of the Social Sciences*, **4**, 1–15. An alternative way of phrasing a somewhat similar idea is that used by Eiseley, who refers to human life as 'surreptitious burning to a purpose' (p. 176), and as the taking over of the oxidation process to produce dreams and poetry (p. 180). Eiseley, L. (1973) *The Unexpected Universe* (Harmondsworth: Penguin). Compare also Dewey, J. 'The more an organism learns, the more it has to learn to keep itself going' cited by Eiseley (1974) *The Night Country* (New York: Garnstone Press), p. 222.
[4] See the discussion in Chapter 5.
[5] See chapter 5.
[6] Thom, R. (Chap. 8, note 29).
[7] For a discussion of the way in which a team of Maxwell's demons might be pressed into service to prevent disasters, see Turner, B. A. (1977) The origins of disaster (note 1).
[8] See Rivas, J. R. and Rudd, D. F. (1975) Man–machine synthesis of a disaster-resistant system, *Operational Research*, **23** (No. 1), 2—21.
[9] These are derived from the categories set out in Chapter 6.
[10] See the discussion in Chapter 6.
[11] See the discussion in Chapter 6.

NOTES

[12] See chapter 9.
[13] See the discussion by Eiseley: Eiseley, L. (1973) *The Unexpected Universe* (Harmondsworth: Penguin), p. 40.
[14] To combine a phrase of Herbert Simon's, with another which, Eiseley informs us, was often attached to Elizabethan legal documents.

Index

Page numbers in italics refer to the notes.

Aberfan disaster   52, 53–5, 57–80 *passim*, 81, 84, 86, 88, 89, 90, 103–4, 105, 153, 178–9, *218, 219, 220, 221, 222, 224*
  birthrate after   *213*
  Borough Council   54, 60, 61, 72, 90
Abnormal loads, 53, 55–6, 60, 66, 74, 150
Accidents, accident reports   14, 49, 81–2, 84–5, 86, 87–8, 105, 164, 173, 177, 178, 181, 182, 183, 187, 192, *212, 215, 217, 223–4, 235, 236, 237*
  and disasters   82, 87–8, 89, 90, 99, 107, 125, 126, 130, 133, 136, 138, 140, 145–6, 149, 151, 154, 157, 158, 162, 167, 170, 171, 180, 181, 183, 184, 186, 187, 189, 193, 194, *225*
Accident studies and safety management   26–30, 31, 84, 87–8, 183–4, *210, 211*. See also: Aircraft, biological, chemical, marine, mining, nuclear and road accidents, boiler explosions, explosions, fire and structural collapse; See also *Sea Gem*, Bethnal Green, Ibrox Park
Administration, administrators   4, 75–80, 150, 151, 153, 163–71 *passim*, 177, 181, 192–3, 201, *210, 229, 234*
  'administrative man'   165, 167, *234*. See also, Organization, Management
  administrative rationality   133–6
Advisory Committee on Major Hazards   26, 29

Aetiological classification
  of accidents   184–5
  of disasters   12–14, 190
Age criteria   122
Aircraft accidents   13, 96, *225*
  1975 Paris DC10 crash   14–15
  Comet airliner failures   23–4
Algeo, Miss   111–17
America, American   10, 36, 37, 132, 143, 178, 196, *216–17, 224, 232*.
  'American Soldier' research   *214*
Amplifying properties of organizations   179–80, 185, 187
Anomalies   150–1, 155–6, 158, 159, 193
Anti-tasks   179–81, 185, 186–7.
Anxiety   42, 43, 44, 69–70, 93, 133
Anzio I, motor vessel   172–7 *passim 235*
Architects   25, 56–7, 59, 74, 176, 177, 178
Assumptions, shared assumptions   120–4, 128, 146, 151, 161, 165, 166–71, 174, 177, 179, 200–1, *211, 228, 229, 235*. See also expectations, norms.
  false assumptions   6, 86, 100–5, 120, 178, 200. See also Error.
Attention   132. See also Perception
Authority, authorities   106, 118, 122, 152, 164. See also Power
Avalanches   13, 190, *206*

Batch-production scheduling   50–1.
Beliefs   84, 85, 86, 99, 102–4, 118, 158, 163, 164, 166

INDEX

Beliefs *contd*
   rigidities of   58–9, 75, 78, 80, 86, 102–4, 133
Bell, Dr. Joseph   223
Bethnal Green disaster   *233*
'Big Dish' project   17–18
Binet   *232*
Biology   127, 186
   biological accidents   13, 28, 84, 108–9, 110–17, 190. *See also* Contaminated fluids, Smallpox
   mutation   *237*
'bits' (of information *q.v.*)   *229*
Boiler explosions   21, *225*
'Bolt-from-the-blue' hypothesis   33, 48
Boston Weather Bureau   40
Boundaries   105–125 *passim*, 197, *226*
Bounded decision zones   168–71, 187, 200, *232, 235*
Bounded rationality, limited rationality   133–6, 138, 161, 164, 167, 169–71, 187, 200
   Principle of   133–6, *228*
Bridges Engineering Design Standards Division, British Rail   55, 60, 66
Britain   10, 57, 58, 81, 97, 201, *232*
British Government sources on accidents   49, 107, *225*
British Fern, oil tanker   119, *225–6*
British Rail   53, 55–6, 60, 61, 63, 64, 66, 71, 72, 73, 104
Bruno, Mr   110–17
Building Regulations   173, 176, 177

Calamity   82, 153. *See also* Disaster
Cambrian Colliery explosion   92–98, 158, *222, 224*
Cambridge   150
Canada   10
Cardiff   118
Cataclysm   82. *See also* Disaster
Catastrophe, catastrophic change   82, 83, 90, 92, 178, 213. *See also* Disaster
   'catastrophe'   150–1, 156, 159. *See also* 'anomalies',
'serendipities'
Catastrophe Theory   153–9, 185–7, 194, *224, 232*
   'fold' catastrophe   154–6
Categories for information   139–49 *passim*, 150, 170, 198, *225, 229, 231, 238*
Cathexis   *221*
Centralization   5
Certainty   135–46 *passim*, *229*. *See also* Uncertainty
Chemical accidents   1–2, 13, 29, 190, 194, *217*
China   9
Civil defence   *214*
Civil disturbance and riots   15
Civil Service, government employees   64, 176, 178
Claustrophobia   69
Cliques   122
Cognitive dissonance   158
Coldharbour Hospital Fire   92, *222, 224*
Collective behaviour   *233*
Collective consciousness   177
Collective stress situations   216
Collisions   118–20, 122, 123, 124, 190, 197, *214*
Command papers   *225*
Commission on Safety in Mines, 1938   55, 58
Commitment   132
Commonsense   85
Communication   5, 52, 57, 61, 77, 107, 126, 164, 182, 186, 197, 210, 220
   codes   120, 124, 139–49 *passim*, 197
   inter- and intra-party   120–1, 197. *See also* Information, Communication networks, Variable disjunction of information
Communication channels   139–49 *passim*, 154–9 *passim*, 198, 199, *231*
   routine   109–118, 120, 121, 124, 196–7
   non-routine   111–18, 196–7

Communication channels *contd.*
  capacity of   139–46 *passim. See also* Observation channels, Information flows
Communication Networks   105–25 *passim*, 160, 196–7, 200
  institutionalized   124–5, 196–7
Community   91–2, 122, 193, *223*
Computer   194–5, *234*
Complaints   61, 76, 78, 87, 100, 102–4
Conan Doyle, Sir Arthur   *223*
Concealment of information   118, *226*
Concord, U.S.A.   143
Concrete system, 'sites' (*q.v.*)
  as   68–9, 73
Conflicts   73, 166, *226*
Confrontations   152, *232*
Consciousness   127–8
Contaminated fluids incident, Plymouth General Hospital   108–9, 110, 123, 180–1, *225*
Context markers   157
Contingent universe   201
Convergence of sightseers at disasters   9, 37
'Corruption' of plan   *236*
Costs   135
Creativity   *236*
Crinan canal   173
'Crisis' decision-making   29–30, *213*
Crowd behaviour   *216, 233*
Culture, cultural factors   37–8, 45, 58–9, 83, 105, 120–5 *passim*, 132, 164, 166, 168, 187
  cultural collapse or disruption (*q.v.*)   86, 89, 96, 193
  cultural lag in precautions   100, 101–5, 125, 133
  cultural readjustment after disaster   85, 90, 91, 92, 95–6, 97, 99, 103
'Cunning adversary', *217, 226*

Danger, hazards   107, 108, 109, 110, 112, 114, 117, 121, 122, 123, 124, 135, 151–3, 193, 196, 197, 198, 199, 200, *212, 224, 233*
  danger outcomes   96
  denial of   44, 60–7, 73, 74, 76, 79, *221*
  minimizing of   71–4, 76, 87, 100, 102–5, 120
  pressures to minimize   73, 76
  confidence about handling   117
  failure to call for help   73–4
  remote dangers   108, 109, 114
Darlow, Dr   90
Darwin   *230–1*
De Tocqueville   *224*
De Lorka Mr   62
Deaths, in disasters   16, 82, 83, 84, 90, 92, 95, 111, 114, 183, 196, *211, 222*
  as determinant of 'serious' disaster   9–12, 16
  corpse disposal   16
'*Decalages*'   *233*
Decision-making   2–3, 5, 6, 29, 54–5, 59, 85, 120–5, 134–6, 149, 152, 160–71, *passim*, 173, 177, 179, 180, 187–8, 200, 201, *212, 228, 229, 234*
  decision strategies   134–6, 161–2, 164, *228. See also* Bounded decision zones, Risk, Decision premises
Decision premises   151, 165–71 *passim*, 174–8 *passim, 228, 234*
  revision of   174–8 *passim, 225. See also* Assumptions, World View
'Decoy' problem   59–61, 64, 78, 80, 86, 100, 102–4, 196, *226*
Definition of disaster   9–10, 19, 26, 81–4, *207*
  illustrated   84–98, 193, *223*
Demolition   117
'Demons', Maxwell's   *237–8*
Department of Health   108–10
Design   56–7, 171–9 *passim*, 195. *See also* Engineering design
  of jet engine   *210*
  of 'sites'   68–70, 76

INDEX

Design *contd.*
  of subways and lifts 69–70
Detective novels and disaster *223*
Devonport 108–9, 110, 123
Differential awareness of
    hazard 151–3
'Diffused' disasters 35
Dioxins 1
Disaster
  accounts of 8–11
  after-effects of 4–5, *214, 223*
  and anti-tasks 179–81
  and catastrophe theory 154–9
  and information 138–59, 188
  and organizations 160–88 *passim*
  and surprise 138–59
  and rationality 160–88 *passim*
  as social pathology 34–7
  classification of 12–15, 183–4, 190
  'decomposition' of 179–81, 187
  definition of 9, 12, 26, 81–4, 84–98, *207, 223*
  development of 84–92, 92–7, 100–5, 106, 124, 125, 138, 160, 181–9 *passim*, *214, 216*
  economic effects of 12
  focalized/diffused 35
  incidence of 1–7 *passim* incubation (*q.v.*) of 81–98
  instantaneous/progressive 35
  lists of 8–15, *207*
  man-made 13, 14, 190, *207*
  multiple causes of 14, 23–4, 75–6
  natural 13, 14, 37, 182, 190
  'need for' 38
  need for multidisciplinary approach to 31–2, 38
  novels and films of 9, *206–7*
  origins of 189–201 *passim*
  possible lessons from 76–80
  prevention of 194–9
  previous studies of 8–48 *passim*
    post-disaster bias of 16, 35–41, 48, 83, *214, 216*
    theoretical assumptions of 39–41
  reasons for studying 1–7 *passim*
  scale of 2, 6–7

syndrome 45. *See also* Accidents, Catastrophe, Energy, Failures, Incubation, Information, Medical problems, Order, Organizations, Origins of, Prevention of, Rationality, Surprise, Theory
Disclosure 109, 112–14, 115, 117, 118, 120, 125, 126, 185
  voluntary inhibition of 117, 118
Discrepant events in incubation period, unnoticed
  events 86–8, 89, 90, 122, 125, 151, 154, 184, 193–9 *passim*, *232, 238*
  types of 86–7, 99–105
Disruption 126–37, 189, 193, 197, *225*. *See also* Order
Distribution processes 108–9, 110, 123, 180–1
Diversity, variety 6, 163–4, 166, 187
Division of labour, distribution of labour 121–2, 129
Dixon, Mr 62
DNA 2
Dock-work *226*
Dortmund 58
Douglas, Isle of Man 56. *See also* Summerland
Dreams *237, 238*
Droughts 13
Dudgeon's Wharf explosion, Isle of Dogs 117, 123, 124, *225*

Earthquakes and tremors 13, 14, 16, 37, 86, 183, *210*
East Bengal cyclone 16
East Ham, London 172
Ecological balance 128
Edsall, Prof. 113, 115
Efpha, motor vessel 119, *225*
Efthycosta II, motor vessel 118, *225*
Electrical fault in colliery explosion 94, 96
Electronics companies 121
Emotion 136, 151, 152, 178, *221*
  emotional factors impeding communication 42–3, 44, 45,

Emotion *contd.*
    46, 77. *See also* Information difficulties
  emotional or affective properties of safety equipment   69–70
Energy and disaster   1, 3, 4, 6, 7, 51, 149, 160, 183, 190, *237*
  and information   3, 6, 182–7, 189–201 *passim*, *237–8*
Engineers   3, 5, 17, 20–3, 54, 59, 62, 139
  and failures   17–26 *passim*, 31, 48, *210*
and rationality   20, 24
  training of   24–6, 173, 175, 176, 178
Engineering design   20–4, 25–6. *See also* Design
  of flats   176–7
  of jet engine   *210*
  of 'sites'   68–70, 76
  of subways and lifts   69–70
  'size effect' in   22, 69–70
English Electric   60
Entropy   127–8, 186, 190, *237*. *See also* Negentropy, Order
Environment   2, 5–6, 14, 128, 129, 131, 149, 161, 168, 170, 174, 187, 190, 193, 198, 199, 200, 201, *206, 235*
  'munificence of'   128, 201
Epidemics   13, 82, 84, 190, *223*
  epidemiological approach to disasters   12, 16, 31–2, 38, 41, 83
Epistemology   *231*
Error   6–7, 14, 17, 46, 57, 87, 89, 90, 94, 99–125, 132, 138, 141, 142, 160, 161, 163, 165, 167, 168, 170, 178, 179, 180, 181, 182, 184, 187, 193, 194, 200, *224, 229, 236, 237*. *See also* Failures, Human error, Unintended consequences
  errors per accident   90
  orderly errors   180–1
  speech errors   180, *236, 237*
  trial and error   168
  typing errors   *236*

Esso Ipswich, motor tanker   118, *225*
Ethnomethodology   *228, 234*
Etiological classification, *see* Aetiological
Evans Medical Co. Ltd., Speke   108–9, 110, 123
Exits   69
Expectations, shared   120–5, *passim*, 169, 181, 185, 187
  framework of   155–6. *See also* Roles, Norms
Explanations, *see* theories
Exploratory trials   135, 167
Explosions   1, 13, 15, 44, 89, 91, 122, 124, 190, 199
  boiler explosions   21, *225*
  Cambrian Colliery   92–8
  Dudgeons Wharf   117, 124
  fireworks factory, Houston   36–7
  Flixborough   29, *222*
  Halifax, Nova Scotia   34–5
  nuclear   35
  Ronan Point   171–3, 176–7

Factory owners   152
'Fail-safe'   173
Failures   4, 30, 105, 121, 128, 133, 138, 170, 171, 181, *209, 236*
  of communication   100–1. *See also* Information difficulties
  of control   7, 70, 191
  of control on controls,   70
  of foresight   50, 77, 92, 99, 107, 161, 170, 179, *217*
  of intelligence   52
  of interpretation   119, 145
  of learning   *236*
  of mergers   *213*
  of problem-solving   56, 65–6, 77–8
  of rationality   5, 136, 164, 173, 174
  of submission   70
  to call for help   73–4
  to respond to warning   87–8
Failures in physical systems   17–26 *passim, 209, 210*
Famines   9

INDEX

Fascination with disasters   9, 17
Fatigue failures   22
Fear   69, 74, 163
Feedback   45–6, 47, 194, *231, 236*
   deviation amplifying   *236*
Fire, and burns   1, 13, 15, 74, 82, 89–92, 104, 181, 182, 190, *207, 237, 238*
   alarm   74
   Brigade   92, 104, 117, 123
   at Cocoanut Grove Club   28, 178, *236*. See also Summerland, Coldharbour, Michael Colliery.
Fire Precautions Act   28
Fishing vessel wrecks   *225*
Flats, see Ronan Point
Fleck, L.   *231*
Flixborough explosion   29, *222*
Floods and sea surges   13, 14, 16, 37, 44, *207*
'Focalized' disasters   35
Fog   119, 123
'Folkways' of safety   85
Foresight, see Failure of foresight
'Forgiveness of system'   19–20
Formal structures/groups   122, 164, *229*
Fortune and misfortune   193, 201, *207*
Frame analysis, frame shifts   157, *224*
Frankenstein   *237*
Functionalist theories   35, 37, 83, 85, 131. See also Theories.

Gall, Mr. William   108, 110
Game   119, 226
Gas, chlorine   1, *212*
   firedamp (methane)   93–8 *passim 224*
   molecules   *237–8*
   nickel carbonyl   *209*
   at Ronan Point   172–8 *passim*, 183, 195
Genetic manipulation   *232, 233*
Geology   *210*
Germany   181

Gestalt shifts   193. *See also* transformations
Gidley, M.   *230*
Global rationality   133, 174
Goals/ends of systems   19, 23, 24–5, 52, 127–8, 132, 134, 136, 138, 161, 164, 168, 183, 191–2, 200–1, *228, 235*
   shifts in   19–20, 75, 101
Grant, Dr.   113
'Grapevine'   122
Groups   42–3, 52, 84, 92, 120–4, 136, 167, 168, 171, 178, 186, 193, 196
   expectations   83, 120, 152, 153, 192–3
   work-groups   120–2

Hail and snow storms   13
Halifax, Nova Scotia
   explosion   34–5
Harding, Mr.   62
Hazard, see Danger
Hazard analysis/audit   30, 78, *212*
Health and Safety at Work Act   28
Health and Safety Commission   26
Helmholtz   *230–1*
Hickson, Professor D.   235
Hierarchy, hierarchical levels   135, 136–7, 145–6, 149, 150, 153, 157, 161, 163, 167, 182, 187, 200, *229, 236*
   'Planful hierarchy'   179–81 *passim, 236*
   of decision-makers, decision premises   161–71 *passim*, 174–8, 179–81, 187
   hierarchical restructuring   158–9
High quality intelligence   6, 7, 51, 77, *206, 216–17*. See also Information.
Hindsight   100, 162, 173, *234*
Hiroshima   35
Hixon level crossing accident   52, 53, 55–80 *passim*, 81, 84, 86–89, 104, 105, 150, *218–222, 232*
H.M. Inspectors of Mines and Quarries   53, 55, 58
Hollingdale, Dr.   113

Home Office   63, 64
'Honeywood File'   25
Hope   128, 163, 201. *See also* vulnerability.
Horse-racing, information games in   *226*
Hospitals   15, 28
Houghton Main Colliery disaster   *234*
House of Commons papers   *225*
Houston, Texas, explosion at   36–7
'Human error'   57, 180
Human problems after disaster   4–5, 12

Ibrox Park disaster   *216, 233*
Ideal type   *222*
Ideologies   *221*
Ignorance   22, 23, 70, 79, 131, 133, 136–7, 162, 198, 199, *229*
  collective   137, *229*
Impacts, *see* Collisions
Ill-structured, ill-defined problems   52, 53, 60, 61, 63, 64, 71, 75–80, 106, *218*
Illusion of centrality   45
Imo, Belgian ship   34
Impact, *see* Collisions
Impact, moment of in disasters   33, 34, 35, 37–8, 40, 41, 48, 216. *See also* Onset
Imperial Chemical Industries   29, *212*
'Incrementalism'   25, 135, 228
Incubation period of disasters   81–98, 99, 100–1, 106, 107, 114, 118, 124, 125, 126, 131, 133, 193–4, 198, 200
  incubation network   88, 93, 105, 194, 199
India   9
Individualistic emphasis of studies and accounts   14–15, 26–7, 31
Industrial accidents   26–8, 152, *211, 235*
Industrial society   130
Influence zones   *235. See also* Bounded decision zones

Informal structure/groups   116, 121–2
Information   3, 20–3, 51, 56, 92, 101, 105, 126, 128, 129, 131, 132–3, 170, 179. *See also* Communication
  and disaster   3, 6, 7, 38, 51, 81, 85, 105, 138–59 *passim*, 160, 162, 165–6, 189–201, *226, 232, 238*
  and energy   182–7, 189–201 *passim*
  content   138–46, *passim*
  definition of   140–1
  discontinuities   146–9, 153–9 *passim*, 193–4, 233
  'fixed'   145–6, *230*
  flows   110, 114, 117, 118, 120, 123, 124–5, 161, 198, 199, 200, *235. See also* Communication channels
  gathering/processing   134, 164, 192, *234*
  measurement of   138–9, 146, 158, *229, 230, 231*
  nature of   138–46
  'ontic status' of   149, *238*
  Shannonian   146, *229, 230, 231*
  theory, theorists   5, 127, 138–46 *passim*, 155, 158–9, *230, 231*
  threats and warnings as   42–3, 45–6
  transmission of   139–46 *passim, 231*
  See also High-quality intelligence, Information difficulties
Information difficulties   56, 61–7, 72, 75, 77–8, 80, 87, 100–1, 103–4, 106–25, 160, 195–9, 225. *See also* Information, Variable disjunction of information
  all information sent   63, 65, 77
  ambiguities   61–2, 64–5, 72, 75, 77, 89, 100–1, *226*, 228
  communication 'for information'   65, 77
  concealment   118, 197–8

INDEX

Information difficulties *contd.*
  distortions 42, 46, 64, 77
  emotional aspects 42–3, 44, 45, 46, 64, 77
  excessive information, noise 42, 51, 65, 77, 87, 103–4, 141, 190–1, 196, *217*
  in complex 'sites' 61–2
  miscalculation 119
  overload, pressure of work (*q.v.*) 64, 87, 101
  over-reliance on informal networks 64, 77
  passive administrative stance 56, 65
  poor communication 64, 104
  presented at moment of crisis 65
  sent to wrong people 64, 77
  vague orders 62, 63, 104
  with strangers 67, 68
  wrong information sent 64, 77
Injuries from falls 183–4, *237*
Inner Isles of Scotland 172–3, 176
'Insight team' of the *Sunday Times* 14
'Instantaneous' disasters 35, 90–1
Institutions, institutional factors 105, 109, 111, 112, 117, 132, 133, 136, 138, 151, 152, 154, 160, 161, 174, 178, 186, 193, 196, 197, 199, 200, 201, *229, 235*. See also Organizations
Intended rationality 129, 160, 171–8, 192, 200
  intendedly rational actions 132–3, 163, 171
Intentions, plans, purpose 1, 4, 126–30, 134, 136, 138, 154, 161, 164, 166, 171, 179, 180, 184, 190–2, 194, 198, *218, 226, 236*
  failures of intentions 18–9, 23, 181, 182, 191, *236, 237, 238*
Interaction 109, 110, 114, 120, 121, 157
Interorganizational relations 75, 178–9, 181, 196
Intuition 11

Iran 11
Isle of Dogs 117
Isle of Gigha, motor vessel 172–7 *passim, 235*

Japan 10, 42, *216, 226*
Journalistic accounts of disasters 8, 14–15, 31, 38, *209, 214, 215*
  of disaster warnings 46, *214*
  exaggerations in 36, *214*

Kin relationships 122
Knowledge, *see* Information

Landslides 13, 190
Language 121, 141
Laplace *230–1*
Latent structure 89, 94. *See also* Incubation period.
Law of Inverse Magnitude 82
League of Red Cross Societies 38
Learning 157, 158, *224, 236, 238*
Leominster 60, 66, 71, 72
Level-crossings, *see* Hixon.
Lexington, U.S.A. 143, *230*
Liberia 119
Life, life-forms 127–8, 146, 149, 201, *231, 237, 238*
Local government officials 171, 176, 177, 178
Locust swarms 13, 190
Lofthouse Colliery, inrush *222*
London 84, 90, 91, 108–117, 122, 172, 196–7, *222, 224, 225*
London School of Hygiene and Tropical Medicine 111–17, 123
Lookout at sea 118, 123, 226
'Ludiologist' *227*

McGregor, able seaman *226*
Mackenzie, Dr. 111–15
Magic 129
Magistrates Association 63
Maintenance procedures 173, 175
Managers, executives 56, 78, 121, 151
  and safety 78, *212*
  unrealistic view of safety by top management 63, 66–7, 79

248 MAN-MADE DISASTERS

Manpower 185, 193
Manriding in mines 236
Maps, models 128, 129, 149–51, 161, 184, *227, 232, 233*
Marine wrecks 13, 22, 34–5, 88, 118–24, 171, 172–7, 182, 183, 184, *207, 225, 235, 236*
Markham Colliery, Derbyshire 172–7 *passim, 222, 235*
Marxism 131
Materials properties 20–3, 68–9
Mathematical models 232, 235. See also Catastrophe theory
Maxwell's demons *237–8*
Medical problems of disaster 4–5, 15–17, 31, 48, 109, 111, 195, *207, 214*. See also Epidemiology
triage 15, *207–8*
Meers, Dr. 110
Merger failures 213
Messages 138–49 *passim*, 155, 159, 170, *229, 231*
 meta-message 157
 message redundancy 141
 mid-message shift 147–8
 monitoring of 142, 149, 159
Method of investigation *218, 224, 225*
Michael Colliery fire *218, 222*
Microcultures 167, 171
Mining accidents 9, 88, 171–7, 199, *222, 225, 236*
 Cambrian explosion sequence 92–8
 model based on gold-mining accidents 27, 28, 89, 90, *211*
 See also Lofthouse Colliery, Seafield Colliery, Aberfan disaster, Markham Colliery, Houghton Main Colliery
Ministry of Aircraft Production *218*
Ministry of Transport 53, 55–6, 62, 63, 64, 65, 66, 68, 73, 150
 Railway Inspectorate of 55, 63, 66
Misinformation 183–4, 185, 187, 189–201 *passim*
 Principle relating to 189. See also Information

Misleading publicity 68
Misunderstanding 182, 185. See also Error
Mont Blanc, munitions ship 34
'Mores' of safety 85
Mycological Reference Laboratory, London 111–17, 123
'Mystery crystals' *209*
Myths 128, *237*

Nagasaki 35
Naming of hazards 59
National Academy of Sciences/National Research Council, U.S.A. 10, *214*
N.A.S.A. space programme reliability 18, *208*
National Coal Board 53, 54, 58, 61, 62, 63, 64, 72, 104, 153, 199
National Union of Mineworkers 54
Nature 131, 141, 142, *217, 226*
 as a cunning adversary *217, 226*
Navigation 173–5
'Near-miss' 45, 96, 119, 182
 subjective near-miss 221
Negentropy, negative entropy 127, 128, 145–6, 180, 181, 190–1
 negentropophagy *238*
Nicolaw, motor vessel 118, 123, 124, 198, *225*
'Noisy marbles' problem *212*
Normality, notional normality 84, 85, 93, 96, 184, 193, *223*
Norms, normative prescriptions 84, 85, 91, 99, 102–4, 120, 166, *211*
 of rationality 151, 162, 163, 175–7
 See also Safety precautions
Nova Scotia 172
Nuclear accidents 15, *207, 214*
 energy 183
 engineering industry 2, 18, 96, *212, 232*
 reliability in 18, 29, 30
 explosions 35
 nuclear attack and studies 36, 44–6, 48
Nursing staff 109, 110

INDEX

Observation, channels of 141–9 *passim*, 154–9 *passim*, *231*
Omniscience 191, 194, 200, *228*
Onset of disaster 85, 91, 95, 96, 103
Order and disorder, order-seeking 126–37, 145–6, 149, 166, 180, 183, 186, 187, 190, 191, 201, *237*, *238*
    order-eating beings 191, *238*
    orderly errors 180–1
    unwanted order 190
Organizations 50, 53, 55–7, 61–2, 89, 104, 105, 108, 117, 129, 130, 132, 133, 135, 136, 138, 151, 153, 177, 178, 180, 181, 183, 184–8, 197, 198, *226*, *229*, *234*, *235*, *236*
    and disasters, 3–7 *passim*, 76–7, 80, 120–5, 160–88 *passim*, 199–201
    bias in 58–9, 75, 78, 80, 100, 133
    coordination 163–8
    decoy problems in 60–1, 78, 100
    differentiation 168
    discipline 163
    exclusivity 61, 76, 78, 80, 100
    failure of problem-solving in 65–6, 77–8
    and incubation period (*q.v.*) 87, 100
    and information handling difficulties (*q.v.*) 61–7
    and intelligence 6, 77
    intendedly rational organizations 130, 133, 160, 163–71 *passim*, 187
    organizational memory 167
    perceptual horizons in 59, 75
    organizational rationality 101, 120, 160–71, *passim*, 200
    organizational rhetoric 165
    organizational style 171
    organizational surprise 186
    organizational tasks 101, 105, 120, 121, 164, 166, 180, 185, *226*, *235*
Origins of disaster 1–7 *passim*, 49, 50, 75–80, 106, 136, 189–201
    neglect of 8, 9, 38–9, 48

    classification based on 12–13
    See also Pre-accident phase, Incubation period
Overwind at Markham Colliery 172–77 *passim*

P11 district, P26 face, Cambrian Colliery (*q.v.*) 92–98 *passim*
Paignton 108
Paradigms, paradigm shift 150, 157, *233*
Parents 68
Paxton, Mr. 62
Pay-offs 165, 168
Pearl Harbor 42, 196, *216–17*, *223*
Pentre Seam, Cambrian Colliery (q.v.) 92–98 *passim*
Perception of warnings and threats of danger 19–20, 43, 46, 55, 66, 83, 89, 106, 120, 132–3, 152, 166, 169, 185–7, 196, *224*, *235*
    jumps in perception 157–8
    'perceptual horizons' 59
    perceptual inertia 154
    rigidities of perception and belief 58–9, 75, 78, 80, 100, 133, 178–9
Perfect competition 5, 6
Phase model of disasters, *see* Development of disasters
Philosophy of tidiness 145, 201
Physiological errors 180, 194, 195, *236*, *237*
    models 128
Piaget *232*, *233*
Pitfalls 25, 76
Plans, *see* intention
Plans to cope with disaster 26–30, 191, *207*, *216*
Planful behaviour, 'Planful hierarchy' 179–81 *passim*, *236*
Plymouth General Hospital 108–9, 110
Poetry *237*, *238*
Police 53, 55–6, 60, 63, 104
Political factors 132, *217*. *See also* Power

Polybrominated biphenyl (PBB) 1–2
Population and disasters 1, 4, 14
Porth Central Mine Rescue
    Station 95
Post hoc experience, *see* Hindsight
Power 4, 124–5, 152, 164, 167, 171,
    180, 187, 191, 196, *211*
  differential power distribution 152
Practice, good, poor, codes of 85,
    92, 94, 96, 102, 118, 151, 164,
    192
Pre-accident phase 27, 153
Precipitating event or incident 85,
    89–91, 95, 103, 107, 112,
    122, 150, 155–6, 193, 198,
    224, 231–2
Preconditions of disaster 31, 75–80,
    81, 99, 122, 160, 188, 189,
    198, *216*. See also Incubation
    period, Origins of disaster
Prediction, forecasting 132–3, 171,
    210, 212, 227
Pressure groups 152
Pressure of work 64, 87, 92, 94,
    101, 196
Prestige 125
Prevention of disasters 194–9,
    200–1, *238*
Primary contacts 112, 117
Probabilities 149–46, *passim*, 161,
    162, 169, 181, *229, 230, 231*
Problems 164
'Prodromal phase' 214
Professional occupations 213
'Progressive' disasters 35, 90–1
Prometheus myth 6
Protestantism *233*
Psychoanalytic view of response to
    disasters 43–6, *221*
Psychological factors 5, 105, 158,
    195, *233, 234, 236, 237*
  aspects of adjustment to disaster
    91, 157. *See also* Victims
  psychological contract 165
Public and hazards 29, 30, 31, 44–5,
    56, 60, 72, 161, 199–200
  public as strangers 67–8, 76, 104
Public Health problems in
    disaster 16

Public Health Laboratory 108, 110,
    123
Public inquiries, courts, tribunals 14,
    17, 24, 49, 50–80 *passim*,
    81–100 *passim*, 107–25, 126,
    150, 171–181 *passim*, 187,
    198, 200, *219, 225, 236*
  as social mechanism 201
  alternative interpretations of 218
'Pulsar' star 150, *231–2*
Purpose, *see* Intention

Quality control 23, *209*
Queen Elizabeth II 210

Radar 119
Radioactive material 2
Radio astronomy 150
'Radius of foresight', 'of action' 130
Railway accidents 9, 13, 21, *225*
Randomness 127, 145, 150,
    179–182, 190–1, *231, 236,
    237*
Rare events 230. *See also* Risk
Rationalist thought *236*
Rationality 5, 7, 21, 126, 129, 130,
    134, 136, 151, 160–88
    passim, 200, 201, *226, 228,
    229, 234, 235*
  departures from 172–7, 179
  functional and substantive
    129–131, 133, 136
  rational maps 129
  and engineers 20
  See also Intentions, Administrative
    rationality, Bounded
    rationality
Rationalization 72, 172, *209, 221*
Recommendations of inquiries after
    disasters 74–5, 85, 91–2,
    96–7, 102–4, 173–7, 187,
    189, 200–1. *See also* Public
    inquiries
Redthorn, motor vessel 119, *225*
Reductionism 145
Reed, Col. 63, 73
Reilly, Mr. Michael 108, 110
Reliability engineering 18, 23,
    17–26 *passim, 208*

Religious affiliation 122
Reluctance to fear the worst 102–4, 125. *See also* Danger, denial of, minimizing of
Repression *221*
Rescue, relief, salvage and recuperation after disasters 4–5, 35–40, 48, 85, 91, 95, 96, 103
  in complex systems 43
  warning emergency services 47
Research 96, 106, *206, 214, 216, 224, 235*
Resources, resource flows 105, 106, 124, 128, 129, 160, 180, 182
Responsibilities 61–2, 63, 73, 77, 92, 101, 102, 130
Restoration of social structure *213, 216*
Revere, Paul 143–4, *230*
Rhetoric 165, *234*
Risk 29, 76–7, *212*
River Loddon Viaduct, falsework collapse *222*
Road accidents 13, *207*
Road Hauliers Association 63
Robens Report on Health and Safety at Work 27
Roles 75, 101, 120–5, 165, 178, *234*
Roman Catholicism *233*
Ronan Point flats, collapse of 172–8 passim, 183, *235, 236*
Rondle, Dr. 113–15
Royal Society for the Prevention of Accidents 63
Rumours 37, 43

Safety factor 23
Safety management and accident studies 26–30, *212*
Safety precautions and regulations 76–80, 90, 91, 92, 93, 117, 118, 123, 152, 193, 198, 199
  attitudes to 37, 44, 46, 54–5, 63, 79
  changes in 96, 102–4
  collapse of 84
  collective neglect of 54–5

culturally accepted 84, 86, 91, 199
failure to comply with 70–71, 75, 80, 87, 99, 101–4, 152
  penalties for failure 74
ignorance of 70, 79, 152
inadequate 108
out-of-date 75, 99, 100, 101–4, 125
properties of 68–70
propitiatory element in 44, 71, *221*
safety at work legislation *211*
on 'sites' 68–9
with 'strangers' 68–9
vicious circles with regard to 59, *219*
St. Mary's Hospital, London 113–16, 123
San Francisco 86
'Satisficing' 134, 135
Science, scientists 3, 142, 150, 157, 192, *210, 229, 230, 232*
Scotland 172–3, *216*
Scott-Malden, Mr. 66
Seafield Colliery, fall of roof *222*
Sea Gem accident *218, 222*
'Self-evident hazards' 68
'Self-fulfilling prophecies' 132
'Serendipities' 150–1, 156, 159, 232
Sewage system, orderly properties of 181
Shock 82, 91, 124. *See also* Disaster syndrome
Sightseers 9, 37
Single-line components 173, 175
'Sites' 67–70, 76, 79–80, 101, *220*
  as concrete systems 68–9, 73, *220*
  effects of size on 69–70
  site meetings and visits 61, 68, 73
Skills 210, 236
Skopje 11
Smallpox, 1973 London outbreak 84, 86, 90–1, 109–117, 122–4, 196, *222, 224, 225*
Smoke inhalation *207*
Social class *211*

Social distribution of knowledge about hazards 3, 85, 106, 107, 113, 151–3
Social homogeneity, assumption of 124
Social loss 82
Social relations of production *211*
Social reputation 124
Socialization 164, 166
Sociology, sociologists 5, 33–41, 48, 124, 130, 157, 192, 195, *211, 222, 224*
Sociological aspects of origins of disasters 3, 16–17, 83, 201
Sociological and social science studies of aftermath of disaster 16, 33–48 *passim*, 83, 127
Socio-technical view of disasters 3, 8, 47–8, 89, 187
Sorcerer's apprentice *237*
South Africa 58, *211*
Speke 108
Starcrown motor vessel 172–7 *passim, 235*
Status 125
Stereotyped view of public 68, 76, 100
'Strangers' 67–70, 76, 79–80, 101, 103, 104, *220*
   difficulty of informing or defining 67–8, *220*
   need to limit access of 67–8, 76
   on 'sites' 67, 79–80
   opportunity on 'sites' 68–9
Structural collapse 17–18, 25, 173, 178
   of bridges 17, 55, 60
   of buildings 13, 171–8 *passim*, 190
   See also River Loddon viaduct, Ronan Point
Structured nature of unintended consequences 179–81. See also Anti-tasks
Structures of society 131, *211*
Subcultures, *see* Cultural factors
Success 128, 138, 166, 167, 180
'Summary events' and 'negative summary events' *215, 224*

Summerland Leisure Centre fire 52, 53, 56–81 *passim*, 84, 88, 89, 101, 104, 105, *219–222*
Sunday Times 14
Surprise 82, 83, 86, 90, 96, 124, 126, 138–59, 170, 184, 185, *231, 232, 235*
   differential surprise 159
   normal and higher order surprise 149–51, 169
   organizational surprise 186
Suspicion 115
Sussex 119
Systems 135–6, 141, 142, 145, 158–9, 161, 162, *209, 210, 220, 229, 233*
   disaster-resistant *209*
   of energy and information 184–8
   open 136, 159, 162, 180
   quasi-closed 142
Systems approach *213, 217*
Szilard *231*

'Tacit' knowledge and skills 25, 26
Tailings, tipping of 60, 103–4
Tasks, *see* Organizational tasks
Technical factors affecting disasters 3, 4, 6–7, 47–8, 89, 201
Technical systems 130
Technology 131, 201
Teleology, teleonomy *227*
Tennessee Valley Authority 132
Territory 114, 116, 122, 123
Teviot, motor vessel 119, *225–6*
Theatre Regulations 60, 70, 71, 74, 101, 104
Theories, explanations, modes of accounting in complex situations 43, 51, 62, 63, 91, 150, 151, 154, 157, 198, *232*
Theory of origins of disaster
   absence of 38
   concepts in 32
   need for 2, 3–7, 32, 38
   possible general principles 75–81
   proposed 49
   summarized 189–201
   used 99, 107, 126, 185

Theory of origins of disaster *contd*.
    See also Development of disaster,
        Origins of disaster
Thermodynamics 127–8
    Second Law of *237–8*
Threat 40, 41, 42, 43, 106, 174, *215, 216, 225*
    as potential information input 42–3
    collective reactions to 37, 42–3
    'total organism' response to 43–4
    and warnings (*q.v.*) 41–8
Time 39–41, *226*
Tip Safety Committee 74
Tips, spoil heaps 53–5, 58, 59, 62, 63, 71, 72, 86, 90, 153, 178–9, 199
Tornadoes 13, 37, 44
    in Worcester, Massachusetts 28
    in Yap, Western Caroline Islands 37–8
    warnings of 46
Transformations, transitions 89, 92, 93, 95, 105, 157–9, 182–3, 189, *223–4, 233, 237*. *See also* Precipitating event
    of energy and information systems 184–201
Translations 121
Triage 15
'Triggers' 224
Tsunamis 13
    warnings of 46
Turbulent changes in environment 6
Two-party pattern 118–20, 123, 197, *226*
Tymawr tip slip 60

Uncertainty, uncertainty reduction 139–51 *passim*, 154, 156, 158, 159, 169, 174, 198, 199, 201, 228, 229
    uncertainty increase 150, 154, 158, 198, 199
Underestimating hazards 71–2. *See also* Danger, denial of; minimizing of
Unexpected events 149–51, 153, 171–7 *passim*, 182, 183, 189, 194–5, *232*

Unintended or unwanted consequences 6, 22–3, 84, 91, 126, 130–1, 136–7, 138, 183, 189, 201, *211*
    structured nature of 179–81
    'principle of orderly errors' 180

Vaccinations 111–17
Value-added model *233*
Values 120, 135
Variable disjunction of information, disjunct information 50–2, 61, 101, *217, 225*
Ventilation in Cambrian Colliery 93–8 *passim*
Vested interests 152, 174
'Vicious circles' with regard to safety procedures 59, *219*
Victims of disasters and accidents 4–5, 8, 40–1, 45, 47, 184, 191, 193
    adjustment to disaster 91, 176–8
Virus, pox, smallpox (*q.v.*) 111–17
Volcanic eruptions 13, 183
Vulnerability, invulnerability 44, 46, 117, 118, 128, 196, *221, 235*

Waivers 70–1, 79
War 4–5, 13, 15, 42, 44, 190, *216–18, 226*
    non-conventional 13, 44
    safety precautions in 44
    as stimulus to disaster studies 35–8
Warnings and warning messages or signs 37, 40, 41, 106, 107, 119, 123, 124, 131, 133, 153, 191, 195–9 *passim*, *210, 215, 216, 225*
    ignored 44, 76, 86, 87, 102–4, 123
    model of warning process 47
    position of intermediary passing on 46, 196
    provision of adequate 45–8, 74
    short period of in nuclear attack 38
    response to 44, 46, 47, 87–8, 191
    and threats (*q.v.*) 41–8, 106

Weeds 190–1
Well-structured problems 52, 60, 74, 76, 78, 103, 106, *218*. *See also* Ill-structured problems
Winding gear failure, at Markham Colliery 172–7 *passim*, 183
Windstorms 13, 82
  East Bengal cyclone 16
Wisdom 133, 167
World view 150, 151, 165–71 *passim*, 178, 187, 193, 198, *235*

Wynn's—road haulage company 72

Yap, Western Caroline Islands, typhoons in 37–8

Zones, *see* Bounded decision zones
Zones of interaction 109, 110, 113, 116
Zuckerman, Professor 113

THE WYKEHAM SCIENCE SERIES

1	Elementary Science of Metals	J. W. Martin and R. A. Hull
2	†Neutron Physics	G. E. Bacon and G. R. Noakes
3	†Essentials of Meteorology	D. H. McIntosh, A. S. Thom and V. T. Saunders
4	Nuclear Fusion	H. R. Hulme and A. McB. Collieu
5	Water Waves	N. F. Barber and G. Ghey
6	Gravity and the Earth	A. H. Cook and V. T. Saunders
7	Relativity and High Energy Physics	W. G. V. Rosser and R. K. McCulloch
8	The Method of Science	R. Harré and D. G. F. Eastwood
9	†Introduction to Polymer Science	L. R. G. Treloar and W. F. Archenhold
10	†The Stars: their structure and evolution	R. J. Tayler and A. S. Everest
11	Superconductivity	A. W. B. Taylor and G. R. Noakes
12	Neutrinos	G. M. Lewis and G. A. Wheatley
13	Crystals and X-rays	H. S. Lipson and R. M. Lee
14	†Biological Effects of Radiation	J. E. Coggle and G. R. Noakes
15	Units and Standards for Electromagnetism	P. Vigoureux and R. A. R. Tricker
16	The Inert Gases: Model Systems for Science	B. L. Smith and J. P. Webb
17	Thin Films	K. D. Leaver, B. N. Chapman and H. T. Richards
18	Elementary Experiments with Lasers	G. Wright and G. Foxcroft
19	†Production, Pollution, Protection	W. B. Yapp and M. I. Smith
20	Solid State Electronic Devices	D. V. Morgan, M. J. Howes and J. Sutcliffe
21	Strong Materials	J. W. Martin and R. A. Hull
22	†Elementary Quantum Mechanics	Sir Nevill Mott and M. Berry
23	The Origin of the Chemical Elements	R. J. Tayler and A. S. Everest
24	The Physical Properties of Glass	D. G. Holloway and D. A. Tawney
25	Amphibians	J. F. D. Frazer and O. H. Frazer
26	The Senses of Animals	E. T. Burtt and A. Pringle
27	Temperature Regulation	S. A. Richards and P. S. Fielden
28	†Chemical Engineering in Practice	G. Nonhebel and M. Berry
29	†An Introduction to Electrochemical Science	J. O'M. Bockris, N. Bonciocat, F. Gutmann and M. Berry
30	Vertebrate Hard Tissues	L. B. Halstead and R. Hill
31	†The Astronomical Telescope	B. V. Barlow and A. S. Everest
32	Computers in Biology	J. A. Nelder and R. D. Kime
33	Electron Microscopy and Analysis	P. J. Goodhew and L. E. Cartwright
34	Introduction to Modern Microscopy	H. N. Southworth and R. A. Hull
35	Real Solids and Radiation	A. E. Hughes, D. Pooley and B. Woolnough
36	The Aerospace Environment	T. Beer and M. D. Kucherawy
37	The Liquid Phase	D. H. Trevena and R. J. Cooke
38	†From Single Cells to Plants	E. Thomas, M. R. Davey and J. I. Williams
39	The Control of Technology	D. Elliott and R. Elliott
40	Cosmic Rays	J. G. Wilson and G. E. Perry
41	Global Geology	M. A. Khan and B. Matthews
42	†Running, Walking and Jumping: The science of locomotion	A. I. Dagg and A. James
43	†Geology of the Moon	J. E. Guest, R. Greeley and E. Hay
44	†The Mass Spectrometer	J. R. Majer and M. P. Berry
45	†The Structure of Planets	G. H. A. Cole and W. G. Watton
46	†Images	C. A. Taylor and G. E. Foxcroft
47	†The Covalent Bond	H. S. Pickering
48	†Science with Pocket Calculators	D. Green and J. Lewis
49	†Galaxies: Structure and Evolution	R. J. Tayler and A. S. Everest
50	†Radiochemistry—Theory and Experiment	T. A. H. Peacocke
51	†Science of Navigation	E. W. Anderson
52	†Radioactivity in its historical and social context	E. N. Jenkins and I. Lewis
53	†Man-made Disasters	B. A. Turner

†(Paper and Cloth Editions available.)

THE WYKEHAM ENGINEERING AND TECHNOLOGY SERIES

1	*Frequency Conversion*	J. Thomson, W. E. Turk and M. J. Beesley
2	*Electrical Measuring Instruments*	E. Handscombe
3	*Industrial Radiology Techniques*	R. Halmshaw
4	*Understanding and Measuring Vibrations*	R. H. Wallace
5	*Introduction to Tribology*	J. Halling and W. E. W. Smith

All orders and requests for inspection copies should be sent to the appropriate agents. A list of agents and their territories is given on the verso of the title page of this book.